KEEPING
BRITISH
BIRDS

KEEPING BRITISH BIRDS

AN AVICULTURAL GUIDE TO EUROPEAN SPECIES

Frank Meaden

BLANDFORD

A BLANDFORD BOOK

First published in the UK 1993
by Blandford, a Cassell imprint
Villiers House, 41–47 Strand, London WC2N 5JE

Based upon an updated and revised version of
A Manual of European Bird Keeping first published
by Blandford Press 1979

Copyright © 1979 and 1993 Frank Meaden

All rights reserved. No part of this book may be reproduced
or transmitted in any form or by any
means, electronic or mechanical, including photocopying,
recording or any information storage and retrieval system,
without permission in writing from the copyright
holder and publisher.

Distributed in the United States by
Sterling Publishing Co., Inc.
387 Park Avenue South, New York, NY 10016-8810

Distributed in Australia by
Capricorn Link (Australia) Pty Ltd
P.O. Box 665, Lane Cove, NSW 2066

British Library Cataloguing-in-Publication Data

A catalogue entry for this title is available from
the British Library.

ISBN 0-7137-2388-2

Typeset by ROM-Data Corporation, Falmouth, Cornwall

Printed and bound in Great Britain by Biddles Ltd., Guildford

Contents

Foreword vii
Preface ix

Part One

1. Introduction 3
2. Housing 10
3. Feeding 18
4. Hand-Rearing Young Birds 40
5. 'Meating Off' 44
6. Pairing, Nesting and Nest Sites 47

Part Two

7. Buntings and Finches 67
8. Blackcap, Chiffchaff, Firecrest, Goldcrest, Warblers and Whitethroats 136
9. Flycatchers, Tits, Bearded Reedling and Wren 168
10. Thrushes, 'Chats', Redstarts, Bluethroats, Nightingales, Siberian Rubythroat, Red-flanked Bluetail, Robin, Siberian Blue Robin, Wheaters and Accentors 190
11. Larks, Pipits and Wagtails 221
12. Woodpeckers, Wryneck, Nuthatch, Treecreeper and Wallcreeper 235
13. Starlings, Waxwing, Bee-eater, Golden Oriole, Hoopoe, Roller, Kingfisher and Shrikes 247

14	Cuckoo, Little Grebe, Dipper, Swallow, Martins, Swift, Moorhen, Lapwing, Nightjar, Partridges, Little Ringed Plover and Sparrows	264
15	Corvids, Hawks, Owls and Golden Eagle	284

Index of Common Names	300
Index of Scientific Names	303

Foreword

FOR ANYONE WHO SETS OUT to write a book covering a great range of related items, the resulting work is likely to be one of two types. The first type appears more frequently. Its author probably has a sound and wide-ranging knowledge of the subject, but insufficient time often imposes a restriction on the material which could be included in the book, so the author must turn to other sources. A work of such a kind, when compiled from up-to-date and accurate sources, is a most useful book, but it is often the case that compiling information of this type highlights some rare species for which there may be only one or two available facts. Much research is then required, going further and further back through the literature until the original publication may be discovered, as maybe a report of observations made perhaps a century ago! In such circumstances, the great importance of first-hand observations is all too apparent.

This brings me to the second, and much rarer, type of book: the primary source work. Such a book is based on the broad, personal knowledge of the author and few people have the fortune or persistence to gain the necessary breadth of experience that makes such a work possible. Frank Meaden is one of those happy exceptions, and in this book he offers his own hard-won knowledge.

I have come to know him as a man who has the avicultural equivalent of what is called 'green fingers' when it occurs in gardeners; I suppose 'feathered fingers' would be the avicultural term. Certainly, his great ability to care for birds is one that many people find it difficult to emulate. By pursuing a keen interest in the keeping of native birds over many years, and with a constant determination to learn more, he has gradually built up a personal and practical knowledge of a surprising range of species. I think he shares my feelings that knowledge, in a subject that you care about, is something to be shared. The information that he has gained has been freely imparted over the years to those who have come to him for help. I am very pleased that he is now able to offer it to a much wider audience.

By its very nature, a book of this kind is a very personal thing, and to some extent must be an individual and idiosyncratic account coloured by the range of the author's experience. Should one think that it is open to criticism on this account, any such consideration is surely outweighed by the importance of the original information and observations that become available. Thus, even the potential critic is likely to turn to it as a primary source of information!

I know that Frank Meaden has always been concerned about the way in which birds are kept. I know too that he will share my view that this book should not only help people to keep and breed birds with greater success, but also encourage enthusiasts to provide better conditions and greater care for their birds. If it does this, then all the time and effort will have been well justified.

Colin Harrison

Preface

EVEN AFTER THREE SCORE YEARS AND FIVE of practical bird keeping, I still try to keep an open mind on all aspects of this subject, although I must confess to adhering very firmly to beliefs in certain areas where well-tried methods have a high degree of success. Inevitably, when writing on a technical subject in which there is a vast diversity of opinion and practice, one must write from one's own point of view and practical experience. Some may gain the impression from my comments regarding experiments that I have become hide-bound, adhering rigidly to theories and principles. On the contrary, however, it is my open-minded attitude which often causes problems, for in any experiment I carry out I investigate all implications of the collected data and endeavour to understand unexpected phenomena as well as accepting the overall picture.

It has been suggested that I give a short account of my life, and, through no false modesty, but since the reader will be interested in aviculture, I shall confine the autobiography to the events which explain my interest in the subject. My introduction to bird keeping goes back as far as memory can recall: my father had a seemingly infinite capacity to store knowledge of all living creatures and, since horses would not fit in our garden and too many dogs did not fit in his income, birds proved the favourite. My earliest recollections are of a nest of song thrushes which I hand-reared from a very young age with the assistance of an early-rising grandmother; as a matter of interest, one of these birds was still singing strongly when I went into the army as a young man. Apart from these and other specimens where I had been told 'these are yours', my first very own bird, from choice, was acquired when returning from a fishing trip with my father, a keen angler. As we neared home I saw a flash of yellow and green in front of a car; it failed to emerge from the other side and in the road I found the still form of a cock greenfinch; picking up the 'corpse' I held it carefully until home indoors, when suddenly a slight movement indicated that it was only stunned. This greenfinch never flew strongly again but lived for many years in a cage with a canary hen and they annually produced their hybrids for five or six years; his death did not occur until I was on the point of leaving school.

I continued bird keeping – stimulated by my father's interest – and watching wild birds in Windsor Park and the surrounding Berkshire, Buckinghamshire and Surrey countryside as a child, and then later in life I studied with the intention of becoming a veterinary surgeon. I took over the research side of a firm producing pet foods a few years after demobilisation from the army and, with aviaries gradually extending in a suburban back garden and then in other gardens after moving house, I found my interest never waned.

A true love of the four seasons and all living things was instilled in me from childhood and I developed a constantly searching and enquiring mind. My aims seemed clear enough: whatever country I was visiting or resident in, birds became the subject of study. Eventually I settled for keeping birds of the western Palaearctic regions and trying to discover the factors contributing to success in keeping and breeding these birds – particularly with regard to replacement diets – and endeavouring to produce strains of bird which would thrive and breed under aviary conditions. Over all those years – from the age of five – I have kept numerous varieties, many of them abroad; gradually, with established residence in England, the choice was dictated – I would keep native birds, or at least those of western Europe.

Part Two of this book is a list of those European birds which have formed part of some experiment in either breeding or feeding; I have coded it with a reference relating to success or failure in breeding (see page 65 for details). With dietetical experiments, of course, one has to accept that breeding cannot be successful on every occasion: a lack or an excess of some foods contribute quite easily in such matters and this is why a well-balanced food intake each day is essential. Although variety is important, a change in the additives, rather than the tried and tested basic food, is best, otherwise losses may result with a consequent need for frequent re-stocking.

Many of the rare species mentioned in these pages, such as hoopoes, bee-eaters, wrynecks, Dartford warblers, barred warblers, pine grosbeaks and long-tailed rosefinches, have been presented to me by friends abroad, zoological organisations and universities for the furtherance of my studies. I write this book with the sincere and optimistic intention of contributing to the knowledge of bird keeping, and to enable those who wish to keep and breed native British birds to achieve greater success. I believe this to be important for a number of reasons: the pleasure that can be derived from doing so; the beauty of such creatures; the greater and more intimate knowledge of these birds which can be gained from close association with them. It is vital that, despite increasing urbanisation, people should be able to know and therefore care about bird life;

opportunities to observe and live with birds, which in an urban environment would not be possible because of the effects of poisonous sprays etc., can thus be achieved. The knowledge of bird keeping may assist those whose work involves, or who wish to save, injured birds or deserted nestlings. Finally, of course, we depend on the bird keeper and breeder to save endangered species.

Even though now I am retired I was previously, almost throughout my entire career, involved in animal nutrition and, subsequently, originated and patented Avi-Vite bird foods. Although frequent mention is made of these foods, I would emphasise that this is solely to enlighten any reader's curiosity as to my feeding methods. These preparations are no longer generally available. Also, where I refer to successful and unsuccessful breedings, I do so with the full knowledge that *pro rata* to the number of species kept, the reader could be excused for believing that the figures might have been improved upon. With this in mind, however, I would emphasise that most of the species, subspecies and races listed came into my care initially in connection with my research work in nutrition, as well as for behavioural studies. On many occasions when breeding success was achieved by the birds, it was a welcome bonus over and above any data I had gathered. The connection between nutritional research and breeding pairs of birds may appear tenuous but diets are all-important throughout a bird's life and play a part in all its functions.

Not all of the birds which I kept were originally intended for my own research: some were the subjects of other researchers, who perhaps lacked the facilities or the time to care for them; some had been in my care for months, at times for over-wintering or when all research had been completed, and often there was no intention to breed them. Thus, in some cases, I was merely fortunate that the birds happened to be in my aviaries at just the right time, that is, when they were beginning to think of breeding.

Today, with the short-sighted or blinkered attitudes and archaic outlook of some protectionists, it needs to be proven to one and all that any species of bird will breed in confinement when cared for in the correct manner. True conservationists, such as Philip Brown, who was with the RSPB (Royal Society for the Protection of Birds) for 17 years, 11 as its Secretary, realised long ago that aviculturists with expertise in avian husbandry have much to offer to the conservation of birds in general and to ensuring the existence of endangered species for future generations (Brown, P. *Birds in the Balance*, Andre Deutsch, 1966).

I have been the Honorary Secretary of the ASPEBA (Association for the Study and Propagation of European Birds in Aviaries) since it was formed in 1964. The Association members, past and present, have contributed at least 60 first breedings to avicultural records and I personally have achieved no less than 28 since 1932 in my efforts to prove that any bird, when cared for correctly, will attempt to propagate naturally under controlled conditions.

Frank Meaden
Emneth, Cambridgeshire, 1993

Part One

General information and advice
on choosing birds, housing, feeding
(including insect culture),
hand-rearing, 'meating off',
pairing and nesting.

1
Introduction

ALTHOUGH, IN THE PAST, birds of European species were relatively easy to obtain, the reduction of bird populations as a result of increasing urbanisation and careless use of insecticides has given rise in the UK to a need for strict legislation controlling the catching of live birds. Bird keepers, nowadays, therefore often suffer some frustration in trying to acquire certain protected species. However, ample facilities do exist whereby those who genuinely wish to breed birds may apply for licences permitting them to take, from the wild, limited numbers for the purpose of aviculture. With the possibility of an even greater scarcity of native species, conservation, rather than the pleasure gained from watching birds, is the finest motive to begin bird keeping.

In the UK all birds, with the exception of a few species – at present, for example, corvids and starlings – are legally protected in one way or another. My own work with birds has, in the main, been greatly assisted by the appropriate authorities granting me licences to take wildlings for research work; likewise a number of foreign governments have allowed me permits to have specimens taken in their countries for the same purpose. I am grateful to all who have helped me in this manner as I owe so much of my existing knowledge to their kind co-operation.

Since protected native birds may not be sold indiscriminately, and in order to make allowances for those who breed birds and do hope to dispose of them by sale, the Wildlife and Countryside Act 1981 permits the sale of birds which have been bred in aviaries, provided that, as nestlings, they have had closed rings put on them. A closed ring is a metal ring which can be passed over the foot of the young bird, and will move freely on the shank of the adult specimen, but cannot be removed over the foot once the bird is fully grown. There is, of course, nothing to prevent one from keeping unringed birds if one wishes to have them for pleasure, study or research in one's own aviaries, or merely to give away any surplus bred from them.

Whilst the ringing of young birds is essential in the UK for the person wishing to sell stock, I have not ringed one of my own birds for many years; I have certainly done so for other people, as I do not like to see a novice attempt this job without some prior experienced guidance. Indeed I have never encouraged a newcomer to bird keeping to commence ringing for a number of seasons as it is common for the parent bird, when cleaning the nest, to take exception to the rings as foreign matter and throw the offending bands out, together with the much sought-after young.

The birds in our family aviaries are never ringed because, in large enclosures which are densely planted, a number of accidents can occur; rings can so easily become caught up on a twig of a nest site, a loose piece of wire or even growing shrubs, whereas birds that are housed in cages or indoor flights seldom meet with such accidents. It can be heart-breaking when one has finally succeeded in breeding a few of a much longed-for species to find losses through ringing. It is also a sound general principle to keep interference with breeding stock to a minimum until one knows them intimately.

Keeping any form of livestock is a responsibility and the husbandry involved cannot be shirked or avoided at any time; when the decision is made to keep and breed birds, the onus is upon the keeper to house the stock under sound hygienic conditions and to feed them a reliable diet. Aviaries must be constructed with the welfare of the inmates in mind, and it is a moral responsibility of any bird breeder to keep the stock in the safety of comfortable enclosures. The life-span under such conditions can be increased to three times that of wild individuals and some aviary-bred varieties have now appeared with brighter plumage and increased size. Natural wild species live in constant fear of predators, undergo many hardships and perils during migration, endure hard winters foraging for enough food to survive, and may even face possible death from toxic chemicals whilst resident here or transient when migrating. On the other hand, aviary stock need not, if their cages are up to the required standard, encounter any such risks and should, in time and with increased domesticity, provide ready breeding strains.

Having decided to keep and breed certain birds, one must next consider which group, for example seedeaters or insectivores, or some of each, and then which species or collection of species. Here it would be wise to consider one piece of advice in particular, if in real earnest about breeding birds: success is far more likely if you keep only one pair in an aviary. Certainly some of the common more easily-bred varieties will no doubt breed successfully in a large collection, but so many young never live to leave the nest under this arrangement. It is difficult to justify such waste. Those who pin their hopes on rearing under such conditions must be prepared to face the inherent risks of many losses.

With regard to simplicity of the care involved, most insectivorous birds are easy to keep in good health just as long as they are fed correctly and not over-crowded. Some will require a shed-type shelter attached to their aviary, to which they may retire in cold weather; others need gentle warmth and, probably more important, extra light during the winter months. Due to the vast amount of small live insects required by newly hatched nestlings, insectivorous birds are not 'easy' to breed. It is useless offering maggots to those just out of the egg because the skin is too tough. They require something soft, such as the small green caterpillars of the green tortrix moth and the large cabbage-white butterfly, which are found on oak and cabbage respectively. Even the mealworm can, at this time, be given only in a small size (larger if it is at the white, skin-shedding stage). Incorrect feeding by no means signifies failure in all cases – some birds will break up insects more than others – but it is of great importance that one should know of the numerous suitable insects which can easily be bred in vast quantities. There is the wax-moth, of which there are three sizes. One chooses from the American race and both greater and lesser European races, depending upon the birds being bred. There are locusts and crickets, and many other species of small insects apart from these which can be cultured. Caterpillars, too, can be cultured or supplied commercially. It will help to grow the type of vegetation which encourages greenfly in an enclosure; hops are very good since the underside of the leaves becomes densely infested with these insects, which are relished by seedeater and insectivorous species alike when rearing is in progress. Even a heap of horse manure in an aviary will attract much wild insect life if kept reasonably moist. Bee or wasp grubs are sometimes obtainable and are highly nutritious items to offer. Wood-ant 'eggs', easily collected in certain areas, are one of the finest foods it is possible to obtain.

Success to one person can mean the rearing of one or two stunted young birds; others only feel that success has been achieved when a full brood of young are flying and indiscernible from the parents. Patience and perseverance are essential factors and of the utmost importance. It is always wise to compare notes with other breeders and to study wildling habits. A holiday spent observing a particular species in the wild, and compiling data for reference when later attempting to breed that bird can be an immense help and perhaps influence the outcome of such efforts. Very often I have seen the comparative beginner to bird keeping be successful in a first breeding when older hands have failed; this is most likely due to a keenness to learn, and a willingness to take advice and to try, for example, new methods of feeding. So many of us, even farmers who have kept livestock for many years, have fixed, inflexible ideas.

When the term 'migrant' appears in this book it refers to those birds migrant to north-west Europe. Depending upon the latitude of the area where the bird keeper lives, and the local weather conditions, the housing of birds will vary in both winter and summer. Winter roosting temperatures should not be below 10°C (50°F); feeding time should be extended with artificial lighting. The more delicate varieties, wintering in tropical areas when wild, need temperatures no higher than 15°C (60°F) as long as their feeding periods are lengthened. Excess heat from direct sun rays during the warmer months can be avoided by completely shading the roof with shuttering or climbing vines, or by positioning the aviary beneath a suitable tree. This is, however, rarely necessary where ambient temperatures do not exceed 32°C (90°F) in the shade, but humidity is vital at high temperatures.

There are some birds which one may leave outdoors all winter, provided that they have a dry shed-type shelter to remain in if they so wish. Some nights they will roost outside but, if their food is habitually placed in the frost-free area of the shelter, they will use it more often. There must be some form of illumination, natural or otherwise, to ensure that they can enter, as birds such as we are considering do not normally enter dark or semi-dark shelters. Mention should be made here that, of the species which may be housed in this manner, those allowed the use of outdoor flights during winter are invariably the foremost in spring song and also breeding condition. This difference can at times be quite remarkable to witness: if one keeps a number of the same species, some of which are housed in one manner and others which must stay indoors all the time due to limited suitable housing facilities, one can observe the beneficial effects of the outdoor life in the general condition, the spring moult and the bloom on the plumage. Although the vast majority of migratory birds tend to be treated as semi-tropical during the winter months, the cold seems in fact to have little effect when they are allowed a little extra feeding time by artificial lighting; they will enter a shelter to feed in the light and so very often remain inside near their food supply. The daily exercise out of doors, the rain and even winter sun seem all to play a vital part in their health. Whilst this method of housing is suitable and even desirable in temperate regions for the hardier birds, it should not be attempted in cold regions without considerable experience in bird husbandry.

All native birds, when healthy, at home in their daily surroundings and with adequate food available of the type needed to rear young, will at some time in spring or early summer reach that phase of their life-cycle

Introduction

where it is normal for them to reproduce. The study of birds during this season alone is well worth every penny of their keep for the whole year. Each season will bring with it something of interest, for keeping and breeding or studying birds can be one of the most fascinating and absorbing of pastimes.

In spring one must plan the pairings; there is always a period here of doubt, then pleasure on finding that a pair accept one another and courtship takes place. There is rarely much time spent waiting for results: fit birds, particularly hardbills, will nest quite soon after pairing. All birds at this time should be receiving a richer diet. The extra light of longer days will soon have its effect and males will be heard singing and claiming their territory. A watchful eye on the female of any pair is wise at this time; despite all the precautions imaginable one can occasionally find a hen having trouble with her first eggs owing to chilly nights. If it occurs often, it can be a sign of the bird's being too fat, because of either faulty feeding or lack of exercise; early nesters are normally the ones to suffer from this problem. Dry warmth will normally assist the passing of the egg.

Later, when incubation is almost completed and the time for hatching fast approaching, be especially observant for egg shells to appear on the aviary floor, feeding shelf, or sometimes the furthest point from the nest site; be hopeful that it is the young hatching and not the eggs being thrown from the nest. Very tactful observation is necessary now and can be most rewarding when it allows one to see the female feeding young. During the vital stage of the next few days, the time and care spent in providing suitable rearing foods, and the confirmation that there is a ready acceptance of such foods, tend to involve one very deeply.

Summer gradually turns to autumn and this is the time to take advantage of nature's generosity and collect from the wild harvest of berries, fruits and seeding plant heads, for the moult, although a completely normal and uncomplicated function if the birds are fed correctly, must never be allowed to deteriorate into the malady it can become with bad husbandry.

The breeding of native seedeaters, as long as one remembers that they have not been domesticated to the same degree as, for example, the canary, is a fairly simple matter if they are kept under near-natural conditions in outdoor aviaries with suitable foods and adequate nesting sites and materials. Although there is often a temptation to have mixed collections, they are not conducive to propagating any strain of bird. There have been many times when I have bred quite successfully from goldfinches, siskins and redpolls on the colony system, with only one variety to an aviary and sometimes up to six males and double that number of females being housed in an enclosure of adequate and often generous proportions. Pairs of bullfinches, hawfinches, chaffinches, bramblings and most buntings will invariably produce large quantities of better-quality young when in an enclosure of their own. Quite apart from having to share much needed livefood for rearing – and it disappears in a short while with a number to eat from it – one cannot keep a really good record of parentage under the colony-breeding system. It can be useful for some particular group studies and can be achieved with goldfinches, siskins and redpolls.

Pairs of the other finches listed, and the buntings, are better bred in their own aviaries, and, most important of all, never with others of their own kind. It must be realised that, in the wild, where any bird is even only mildly territorially-minded, it can chase from its selected area another bird which it may feel is interloping. Under aviary conditions this is frequently impossible. Death or any sort of injury in an aviary through fighting for territorial rights is unpleasant and unnecessary and is completely the responsibility of the bird keeper. A little extra thought when planning the pairing can make obvious the risks of housing certain birds together. If in doubt take advice from an experienced bird keeper and always err on the side of caution.

A bowl of insects in the morning may last a pair with young the best part of a day; the same amount in a mixed collection could disappear in an hour or so and, for the remainder of the time until the next feed, what are the young to be fed upon? Under such conditions it is grossly unfair to class the parent birds as bad feeders and blame failure upon them. Assess the birds' needs correctly, provide the necessary, leave them alone and it is very surprising just how often the pair will be successful.

Throughout the year make a note of each item of behavioural interest; even commonplace occurrences can all be entered upon a simple index card. Record not merely the number of eggs or young, but also the date paired, when the site was selected or accepted, nest materials used and even the behaviour of both parents throughout the breeding season. In this way a whole family history can be built up which can often prove very useful in successive years. Where one is genuinely prepared to build up experience, such records will show each bird's characteristics, its likes and dislikes, every success and, when failures occur, not just the failures but frequently an overall picture illustrating the causes of such failures.

An aviary need not be just an obstruction to the view, a frame covered in rather hard-looking wire; with planning many enhance a garden and blend extremely well with their surroundings. The reader will belong to the already converted but a pleasant-looking structure can give everyone a far better impression of bird keeping in general, than can a stark ugly one. The facility for free flight, without too much in the way of perches, encourages the birds to exercise more and the perches, if erected by the owner, should be around the outer perimeter and where shelter is provided. If ample natural growth is provided there is less need to offer such perching arrangements, except beneath the roofed portion where very little will grow. Almost any wood will suffice, beech being very good as it can be cleaned so easily; dowelling or any bought timber should be rejected in preference for natural twigs and branches.

2
Housing

FOR SHEER AESTHETIC PLEASURE, large aviaries of grand proportions, well stocked with interesting subjects, do have much in their favour. This was the custom many years ago among those more affluent people who appreciated the company of birds. Such an enclosure can be very attractive as a show-piece: many in the old days were large conservatories holding a variety of trees, shrubs and plants from distant countries, together with collections of birds. Only rarely, however, were birds stocked with thoughts of preserving the species; conservation was not seen as the important subject it is today. If one has unlimited space at one's disposal, one can still have large enclosures – birds will certainly benefit from ample wing exercise and to some it is essential in advancing breeding condition – but the mixing together of many species, or too many birds in a confined space, is to be frowned upon, for when kept in large numbers the majority of species seldom breed as freely or numerously as they would in the wild or if given an ample-sized aviary to themselves. Today conservation must take precedence over aesthetic considerations.

We must then, if serious in our attempts to breed birds other than the common finches, provide the pairs with individual breeding quarters; some of the finches can be left to reproduce in large numbers, provided that they are in collections where considerable thought has been given to their fellow inmates, the provision of large neutral feeding areas and adequate nesting sites.

Certainly, when contemplating reproduction under controlled conditions, one will find that the vast majority of insectivorous varieties will require their own enclosure. It is not merely territorial instinct which dictates this, but more the adequacy of the right type of food at just the right time, that is, when any young are actually being reared (the critical period is the first 6 days). A mixed collection means shared rations whereas one unit per pair means controlled feeding. It is that simple.

Large walk-through aviaries are very popular where collections are open to the public and I also like them for privately owned collections as the birds do tend gradually to gain a greater confidence. They may be used when colony breeding or, if one has decided to maintain a collection of various species. Such aviaries can be extremely attractive when well planted, but, with regard to aviary vegetation, remember that, whereas with insectivorous birds almost any type of flowering plant, shrub or tree may be utilised to either visual or insect-culturing advantage, seedeaters housed in a like manner would cause excessive damage to growing vegetation. In their aviaries it is far better to grow those items which either provide food in themselves (see Chapter 2), in their berry or fruit growth, or attract the insect life which most birds require when breeding is in progress. A number of feeding points should be provided in an aviary of this type.

The walk-through aviary can be shaped to suit one's preference, the landscaping of the property or the ground available for the purpose; the basic essentials are to choose materials for their durability and to give ample thought to the actual planning beforehand. Remember that discussion with other bird keepers and visits to a number of collections at the planning stage will enable one to benefit usefully from other people's experience.

New or used scaffold tube, timber, solid steel rod of 1 cm (about $\frac{1}{2}$ in) diameter or more, perforated angle and small-bore pipe can all serve very usefully in the construction of aviaries. Bricks, new or secondhand, ornamental stone and building blocks should be considered once the general plan has been decided upon. If you are using any steelwork, even if it was once galvanised, a certain amount of wire brushing will be necessary to remove rust, and it must then be painted with an anti-rust solution. Where a smaller enclosure is contemplated, a visit to a local scrap-metal yard will frequently solve one's constructional problems as many can supply the basic frame materials at low cost.

Sometimes one can effectively combine the use of timber and metal. Timber railway sleepers offered for sale can be sawn lengthwise by a local mill to provide strong well-preserved uprights that will last many years. Positioning these uprights at each multiple of their length, with slightly thinner sections linking them and secured to the upper ends, does form a strong frame for a walk-through; it merely requires covering with wire mesh, and timber boarding to be placed at the outer perimeter at ground level so that the wire can be fastened at the base. The main resulting impression is of a very open-type aviary, suitable for all-year use where resident species are concerned and sufficient evergreen roosting facilities exist for their shelter. Always beware of too large an unsupported area, particularly on the roofing section where wire may tend to sag. Make use of straining wires for support where there is a risk of heavy winter snowfalls and limit unsupported wire roofing to 1 m^2 (3¼ ft^2). The end result can be a complete garden under wire and, when tastefully planted, the wire itself is very unobtrusive. Certainly this type of aviary can provide opportunity for studying one's subjects under pleasant conditions.

There can be disadvantages to this type of aviary. After a good breeding season, one may wish to catch up certain of the inmates; I have spent many a frustrated hour crawling under shrubbery only to see wanted specimens re-emerge in a far corner. It can assist if inner corners are constructed in such a manner that they include a wire enclosure into which birds may be driven; a canvas or net, as used over soft fruit, can then be dropped over the exit and catching be made simpler. Even a feeding area incorporating four wire doors, which are usually held close to the roof but which may be lowered to take wanted specimens, can prove helpful on such occasions.

This type of aviary is not suitable for more delicate migrant species which have to be caught annually. The experienced bird keeper would not use an extremely large walk-through type enclosure for migratory birds. There are many reasons, including the facts that it is essential to catch up birds at times and that it is difficult, unless a shelter is provided – somewhere into which the birds freely enter for food and so can be held during the colder months, to keep a close check on stock which could be at a high risk.

This kind of aviary is altogether more appropriate for colony breeding of seedeaters if some study is being made of them as a group, or for a few insectivorous birds that may be kept and bred in close harmony: waxwings will breed more effectively in such a manner; redwings too give little cause for concern when in a family number; hoopoes and even bee-eaters will be happy in colonies.

Few would use such a large enclosure for only one pair of birds unless a serious attempt was being made for a first breeding of some importance to research. One can house a few wisely selected mixed pairs of resident insectivorous species in such large aviaries; a number of neutral feeding areas must be provided, along with much growing vegetation, such as shrubs and small trees, pampas grass, raspberry canes, currant bushes, heathers, brooms, hops and very thick undergrowth in corners and outer perimeters. Nest boxes of a wide variety of types must be made available in abundance and perhaps a floral assortment of shrubs decorating the central area will assist in providing territories. All this adds up to a highly interesting pastime for the bird keeper but it must be said that the breeding results in such aviaries are frequently rather disappointing.

Even if a wide range of cover is provided to encourage insect breeding, and heaps of horse manure or compost are sited here and there, with the consequent occupational insect hunting for the birds, there is often a higher rate of mortality from fighting in these large aviaries. It is completely natural for some birds to select a territory and be prepared to fight to the death to hold it, and one must remain ever alert for any over-aggressive individuals. Do not delay the removal of a bird in such circumstances.

When one decides to keep mixed pairs in this manner, it is of some considerable importance to recall the methods of nesting in the wild and to avoid housing together pairs that may compete for sites. Try a tree-hole nester and a ground-site pair – those which may select a site in low vegetation. When attempting to breed small insectivorous birds there are so many factors playing against success that one must give great consideration to their well-being lest we contribute to their failure.

If the birds are merely for public exhibition in this type of aviary, keeping anything other than the more common varieties is only justifiable if adequate catching points are incorporated in the construction.

Due to the vast area a walk-through-type aviary can involve, it is vital during early planning and construction that one gives sufficient thought to security and to the exclusion of vermin. The use of rubble and concrete foundations all around the outer limits, ideally a path entirely surrounding the enclosure with the inclusion of a few courses of bricks on which to erect the outer frame stands, can help to keep out rats, weasels and even fully grown mice. The entry of very young mice is another problem. They frequently enter when small and grow to adulthood inside, and are thus unable to escape even if they wish to do so; they breed and one then has the family to contend with. The regular use of traps, harmless to birds, is helpful and a wise precaution at all times.

With numerous shapes and types of aviary to choose from, it may be hard to decide which, but do not rush the planning stage. Aim to keep maintenance to a minimum; the renewal of wire, timber and wood preservative treatment must all be considered. They are time-consuming tasks that one invariably finds need doing at the same moment as the aviary is required for breeding birds. Endeavour to construct the aviary so that all panels are interchangeable; in this way one can be removed quickly and a replacement set in position in a matter of minutes. The worn portion needing attention can be concentrated upon later, away from the aviary site, at leisure. Plan according to good ideas seen elsewhere; consider simple methods of feeding in the dry for winter months, built-in observation points, regular supplies of fresh water – all are well worth investigating and can mean much to the person who has other duties to attend to in addition to maintaining avian stock.

A type of aviary rather common in the keeping of parakeets – long flights more or less open to the weather joining at one end a warm shelter in which they feed – can very easily be adapted to the needs of native species, preferably housed in single pairs or perhaps as a limited number. The extra thought and the planning in the early stages of building can mean far more time to enjoy or study the inmates of the enclosures once they are installed and when leisure hours are at a premium.

Considerable success has been achieved using rows of aviaries built in series. When doing so, it is advisable to house totally different species adjacent to one another. In a row of six such aviaries, we could house the following in consecutive enclosures without any undue risk of bickering: bullfinches, accentors, greenfinches, tits, goldfinches, and finally wagtails. In this manner a number of birds can be studied and bred in quite a limited space.

Among other types of aviary there are those of wire on a simple wooden frame, made to be moved from a soiled patch of grass to a new site, perhaps solely for use during the breeding season. More sturdy structures have draught-proof quarters consisting of a shed-like shelter, and some are even attached to a bird room with indoor flights for the inmates' use. I personally favour the row of aviaries as illustrated (Fig. 1), which simplifies feeding and watering during hard weather. The 'party-wall'-type division between each adjoining aviary cuts wire costs and, with one pair to each, it offers complete control. When wishing to do so, one could build it with linking doors so that, in autumn after the breeding season, the row of aviaries could become one, giving greater freedom to stock and allowing maximum flight room.

The initial costs need not be great for aviary erection. Even the simple wooden frame covered in wire, joining four posts top and bottom with boards from post to post, with a door at one corner and a little roofing at each corner to provide the barest of shelters, can be quite attractive when, for example, covered by climbing roses, and it will house a few pairs of breeding birds. This type of enclosure, built into the corner of the garden, takes up very little room and can provide hours of pleasure and educational interest.

My own choice of aviary is shown in Fig. 1. It provides shelter in a feeding and roosting corridor which affords storage space and good observation points. The shelter itself is situated over either a cupboard or open range which comprises storage space. It should have roof lighting over the passageway rather than the shelter and each indoor shelter can have a glazed portion sited above the entrance hole. Beneath

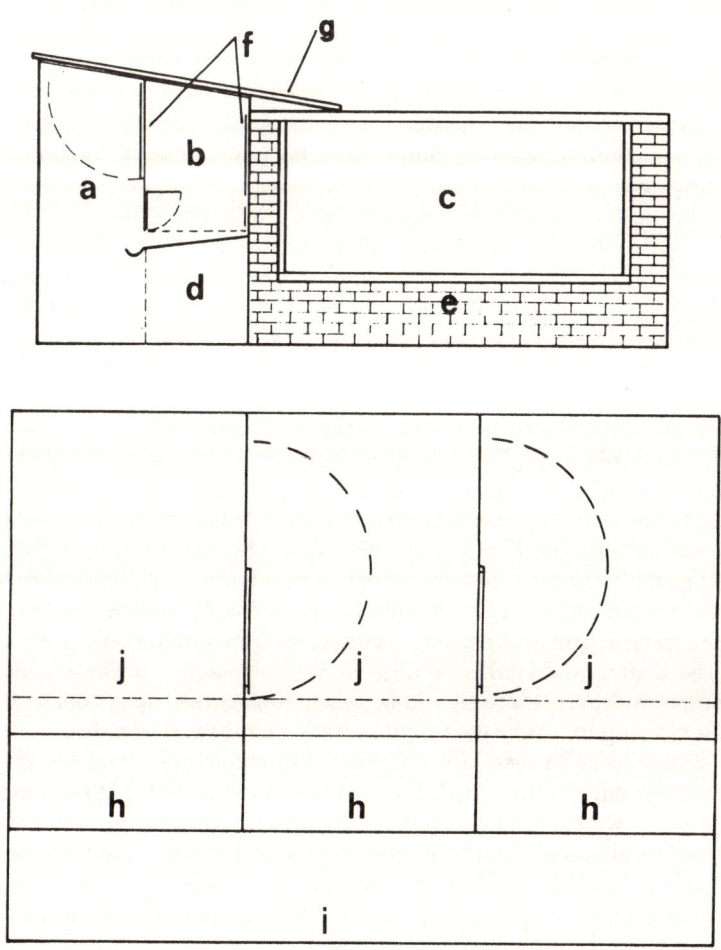

Fig. 1 (*Above*) Side view of a suggested 'in-rows' aviary layout: (a) roofed feeding corridor; (b) shelter with feeding hatch; (c) outdoor aviary or flight; (d) storage space or unit; (e) brick base and piers; (f) optional observation windows; (g) overlapping roof. (*Below*) Plan of the same aviary: (h) three shelters; (i) roofed feeding corridor; (j) three outdoor flights with indication of doors for use in autumn and winter, when aviaries are opened up into one large enclosure. Note that entry into the outdoor flights could be through the rear of the storage space.

the shelter should be a heavy-gauge wire false base sited above a sloping solid base and edged at the corridor with plastic guttering. This assists in catching waste food, droppings and excess water, generally helping to maintain cleanliness; brushing and washing of the wire base remains the main chore. From the feeding corridor I prefer to see a further window or wired section between the birds' shelter and the passageway. Wire is preferable since it allows warm air to circulate in the building as a whole.

The outdoor aviaries are seen to have a low brick surround and brick pillars at each corner; this naturally requires some form of foundation, but, as some sunken concrete is needed anyway as a vermin deterrent, it need not be additionally very expensive, especially if tackled by oneself or a group of interested friends and helpers. The wooden frame set within the pillars for fixing the wire into place can be removable so that simple rewiring operations may be made at a much later date, or they could be fixed in position with bolts set into the brickwork. With such a design the strong old secondhand timbers look very well and, once climbing roses or similar decorative vegetation become fully established, it will blend in with most gardens. Electricity should be laid on by a professional tradesman or advice sought on this. Although rather limited lighting, and possibly heating, may have been in mind originally, make sure the wiring is adequate to cover further heating if insect culturing or the use of an electric drill or other tools is foreseen. If for finches only, the front of the shelter could either be open or have removable sections. If open fronted, however, there should be a generous overhang of roof to stop rain blowing in, to the detriment of the structure and inmates.

Where large aviaries are constructed in temperate climates such as that prevailing in the British Isles, one must remember that the weight of snow in winter can be considerable. Upright posts and cross members should be planned with this in mind and with 13 mm (½in) wire mesh, an effort should be made to ensure that this remains taut at all times. Wire flapping in the breeze, or sagging wire on an otherwise attractive enclosure, can ruin the effect.

Try to have roof lighting in the bird rooms; this will provide the maximum amount of natural light. The only drawback is excessive heat in summer months but, in almost all cases, birds will be in the outdoor part of the establishment at this time. I like to use a bird room for wintering most insectivorous stock but allow birds the daily use of the adjoining outdoor aviaries leading from this. The most delicate specimens I house indoors all the winter.

Adequate thought should be given to installing ventilation in any bird room; an even temperature should be aimed for, but with most species a minimum of discomfort will be experienced as long as the feeding hours are extended. Indoor flights in a bird room are preferable to cages, although it may limit one to a lesser number of birds. However, more species will live amicably together in close proximity during the winter than at any other time, although this must never be used as an excuse to crowd birds in any compartment.

When constructing any aviary or bird room, the incorporation of a safety compartment is always a good investment. This is a method of security whereby one enters through one door, closing it afterwards, before entering the enclosure proper and closing the second door. At times birds can become so tame as to be a nuisance and no doubt, when a bird does fly off, it will be the most valued specimen – it is always so. The use of a safety compartment eliminates such worries and it is well worth the small loss of space it entails. Quite often escapees can easily be coaxed back to an enclosure whilst some have been known to live in the area for years, daily returning for food.

We will later (page 63) look at another way of studying and breeding birds termed 'liberty breeding', a truly fascinating procedure and extremely educational when time can be afforded to study the behaviour of the liberated and returning breeding stock. Frequently, when overstocked with the more common of finches, we have liberated a number then found that we still had to feed them in the garden because of their dependence and stubborn refusal to leave the premises, but of course such trust from birds can give great satisfaction and interest to the bird keeper. I have known birds actually draw blood attempting to get through the wire when endeavouring to return to an aviary in which they were reared. One group of young accentors caused much concern through this, even though they were fully self-supporting and it ended only with transporting them into the country for release.

3

Feeding

WITH SUCH A VARIETY of native European avifauna from which to form a collection, it will be obvious that many different diets may be required. There will be even more to choose from!

The use of any natural or home-prepared food cannot be proven ideal when used over only a short period. Even use throughout the four seasons of the year will not necessarily reveal its deficiencies or, for that matter, where certain items are being supplied too generously. If one has a proven method of feeding and the birds moult in good colour with no long-drawn-out spells of rough feather, winter well with regular short bursts of song, undergo a spring moult normally, pass through it with excellent results and reach breeding condition in early spring, and if successful breeding is achieved and the birds have completed a full cycle on such food, then do not be in a hurry to change. By all means make trial additions of other foods, either insects from home culturing or purchased, or commercially prepared foods which are available in abundance today. We cannot all be experimenting dietitians and spend a life-time pursuing the many varied diets, otherwise so much in other spheres would be neglected. Take advice from more experienced bird keepers who have bred a variety of species consistently over the years.

Generally we feed all seedeaters two mixtures, a soaked-seed mix and a dry mix, varying both the mixtures and the proportions to accommodate individual species' requirements. A soaked-seed mixture is left for 24 hours in clean water and then washed under a running tap and strained. It consists of pigeon conditioning seed (Haiths), extra-small pine nuts, mung beans and safflower, plus smaller quantities of plain canary seed, mixed millets and sunflower. The dry mixture is made up of both British and foreign bird conditioning seed, pinhead oatmeal, teazle, lettuce, chicory and freshly ground peanuts. (Peanuts should be ground daily due to the tendency of this valuable food (25 per cent protein) to develop a dangerous fungal growth if stored too long after it has been ground.) Sunflower becomes a main ingredient of the soaked mixture when birds are rearing nestlings. Where I mention using screenings, I do so in the knowledge that they are a rare commodity today, but one which does seem available on occasions. Without doubt, this dry mixture can be a valuable food, but it is also rich and needs to be used sparingly; a little each day should be given to all seedeaters.

The softfoods which we usually give are Avi-Vite products, but even a human baby food, by itself or mixed with milk plus hard-boiled egg, either whole or just the yolk, will become accepted if served as a thick paste liberally sprinkled with perilla seed. Initially, try embedding the seed in the softfood surface with the thumb. In this way, the birds will soon taste the paste when greedily taking the perilla. Perilla seed (*Perilla frutescens*) is an exceedingly valuable extra to include in the diet; it originated in the Far East and in Japan the leaves of *Perilla frutescens* are used for culinary purposes, in the same way as herbs are used in the UK. Oddly enough, even if they have never seen this seed before, birds go straight to it; even a pinch or so placed at any particular point on the feeding tray is targeted immediately.

Within these pages I refer to my habit of providing soaked seed all year round, even during frosts. It is fully appreciated, however, that many readers experience more severe winters than I do; to those I would suggest that they wean their birds off soaked seed well after the moult and then revert back to it with the departure of winter conditions. Weaning is a simple process: offer a mixture of 50 per cent soaked and 50 per cent dry over a 2-week period, then gradually reduce the amount of soaked.

Feeding for breeding can be vastly different from feeding for exhibition purposes. Weight can be put on any good body bone structure, though not on skull or brow, and this may be required for competitive showing of stock. It is not, however, conducive to longevity, propagation or general well-being; laziness or the inability to exercise the body brings advanced obesity, a shorter life-span and a consequent drop in the production of healthy young to carry on the strain.

Aim for a diet which provides a regulated food intake. Certain birds will endeavour to take their fill of favourite foods to their own detriment; do not allow it if at all possible. If giving a seed mixture, ensure it is well balanced, with not too many fattening seeds; thin it out with clean screenings. Screenings are the wild seeds extracted at source by the farmer at harvest time, when threshing the grain for domestic use, unfortunately they are not very readily available nowadays. Remember that so many seeds available today are completely foreign to our stock; we offer only a substitute diet in all cases. If offering a softfood or insectivorous mixture, make absolutely certain that it is mixed thoroughly, with the additional and plentiful use of greenfoods; vegetable protein is as vital to the health of birds as it is to that of most other creatures. Animal protein as we tend to provide it comes mainly from one source, but consider supplementing this. Imagine an owl feeding off one rat or an equal weight in mice; the rat provides one each of liver, heart and stomach, the latter usually holding a quantity of undigested vegetation of farinaceous items, whereas a number of mice would offer a wider variety and greater number of nutritious organs. Ox heart and liver can be used but a few locusts, wax-moth larvae, mealworms and aphids will provide more variety. Aim always for a wide variety of foods. Trace elements can be missing so easily from a diet and the resulting deficiency may not be noticed for quite a long time. Give thought to what the insects have actually eaten; there is always an amount of undigested food in them which is transferred to the feeding bird.

Variety is just as essential for seedeaters but, with the use of vegetation which has been proven nutritious for human consumption, it is easier to be sure they are being fed correctly; free food, which nature provides and which was used to a far greater extent by our forefathers, is obtainable by all. For the town-dweller it may well mean a trip into the country in autumn for collecting and storing seeding plant heads, but much in the way of natural food can also be grown in a small town garden, tubs, boxes or conservatory. Those in outer suburbia or in the heart of the country have everything at their finger-tips. A library book on wild plants may assist in identification if necessary; the ever-common dandelion, chickweed, sowthistle, comfrey, water-cress and ripened dock seed will be useful. Some may be stored in seed or berry form. Greens such as dandelion and comfrey can be dried for use in softfood during winter months.

To supplement the wild plant life, commercially prepared mixtures are convenient. Foods eaten by birds vary greatly from one part of the world to another; even from district to district there can be vast vegetational differences. Nevertheless, in the wild, most birds can obtain the minimum requirements for a well-balanced diet. Our aim is to provide a similarly nutritious alternative if the birds' health is not to suffer through a build-up or lack of certain elements and this can sometimes happen so gradually as to be undetectable until too late. Our general well-being, health and ability to live as we wish results from the selection of a balanced diet. We are able to achieve this quite easily, both for ourselves and, where big business is concerned, for our stock, whether cows, pigs, horses or dogs. There has been ample research to ensure that their catering needs are met. But who has ever worked out the bare necessities of the Dartford warbler or any other small bird?

Fortunately over the years, more so since the subject of conservation has been taken seriously, bird foods have been improved. Today, with the addition of fresh fruit, vegetable matter and fish or meat protein, a number of good basic diets can be safely relied upon not only to keep adult birds healthy but also to rear nestlings to maturity without signs of rickets or fear of other complaints caused generally by malnutrition. Food must, of course, provide all that is required for full growth, but it must also supply nourishment for the ample maintenance and repair of the body. Each body carries out the wonderful process of turning the food intake into materials and energy essential to its growth, its survival and ability to reproduce. Without going into too much detail, for many chapters could be written on this subject alone, as essential as water and oxygen are carbohydrates, fats, proteins, minerals and vitamins. In some cases excess is passed through the body, in others, too rich an intake can prove harmful. A gradual deficiency can kill, as can excess over a period. It will be seen that the advantage of a varied diet is that, where one item is failing, another can sometimes rectify the balance.

I was asked once to provide details for the Royal Navy of a bird food which could quickly be prepared from readily kept items whilst at sea, and which could be fed to migrant species of birds which frequently land on vessels in an exhausted condition. This can be achieved quite simply, and I include it here for use in an emergency; do remember that it should not be used over any great length of time. It is useful where a damaged bird is handed in for attention without prior warning and when adequate food has not been obtained. Almost any farinaceous matter, such as finely ground grain or flour, wholemeal, maize meal, wheatmeal etc. can form a basis. To this should be added a little soya flour, finely minced dry meat or fish, generous amounts of grated cheese and a little dried milk or liquid if available. Olive oil, soya oil, corn oil and sunflower oil can all be used to dampen the dry mixture; finely minced fruit, such as household currants, sultanas, apricots and prunes, may be added. I have even used sardines (in oil) in emergencies and so many ordinary stored items of human foods can be put to such good use if a little thought is first given to the normal intake of the casualty that there is really never a need for bread sop to be forced down an injured bird's throat. All such foods should end up crumbly moist and sweet-smelling; honey and moist sugar or glucose, apart from their nutritional value, often help to make such an emergency food more palatable.

For regular feeding one cannot do better than to obtain a food from a good, well-recognised source and provide the fresh items, which cannot be stored easily, oneself; remember that young birds require more calcium in the food than an adult bird which will invariably obtain much from food and insect life taken in an aviary. Bird keepers with keen angling friends will be able to provide whole fish or a cleaned-out portion; this can be steamed and mixed into insectivorous foods and provides a valuable addition and much variety to a diet. The use of fresh fish fry, such as small minnows, as a bird food is commended; these, placed in a shallow container, will often be fed by adult birds to their young. Almost everything of redstart size upwards will take advantage of their provision.

The inexperienced bird keeper often makes the mistake of hatching out a multitude of maggots into flies. Unless a fly is fed on honey or glucose water from the time of hatching it has little in the way of nutritional content and should therefore not be fed in any number to birds; the odd few do no harm, but on a steady diet of them, a bird will soon show signs of malnutrition. I have been horrified to hear of how so-and-so's birdroom is full of flies and his birds eat voraciously from them. Inevitably, a few days later one hears of the birds having a membrane closing over the eye and severe eye irritation setting in; if caught in time I have recommended the use of Abidec (Parke, Davis), a vitamin supplement, one drop direct to the beak of a softbill every few hours, combined with the force-feeding of mealworms or wax-moth larvae. However, prevention is always better than cure.

Even among a group as small as the Corvidae the required diets can vary greatly from one bird to another. One can easily understand how a large family such as the warblers (Chapter 8) need to have a widely varied food intake, some needing a far more vegetable-based diet and others extra fruit whilst those such as the small leaf warblers keep in better health when receiving extra milk in their diet to help replace the minute insect life taken in the wild. Milk can, indeed, be used to dampen almost any softfood to the benefit of the subject but, in the case of willow warblers and chiffchaffs, a high-protein invalid food supplement can be mixed with milk and given in small quantities in addition to the basic softfood ration.

One advantage of feeding softfood is that, where it is deemed necessary to include any colouring agent just prior to and during any moult, it is a simple matter to do so. There are many colouring agents to choose from, mostly harmless vegetable-based ones which are used in human foods.

Many softbills will take soaked household fruits such as currants, sultanas, raisins; where it is available, the fruit of the elder can be dried for winter use in this manner and one can freeze many wild fruits and berries collected during the autumn wild harvest; these simply need to be removed from a freezer, in the same way as domestic food, well before use. Apricots, figs, dates, prunes, dried apples and pear may be minced finely and included in the softfood regularly; they are almost all highly nutritious foods which will be appreciated since they are taken quite naturally in the wild by many species.

The inclusion of comfrey, the brassicas, spinach, dandelion (root, leaf, flower and stalk), chickweed, young nettles or even the top couple of centimetres or so of the more adult plant, all chopped finely or minced before use, can be added to the food of any insectivorous species; some will benefit from a more liberal supply than others. All bird breeders who keep hardbills should include in their quarters growing comfrey, cabbage, spinach, dandelion etc. Keep a supply steady during the whole growing season. The remainder of the time, during the winter months, winter cabbages or Brussels sprouts may be left for them to eat from. Old cabbage stalks have always been planted in my enclosures as they invariably flower and seed and the pods are avidly taken by almost all finches while they are rearing young.

Hops may provide only a limited harvest of seeds and over only a short period of the year, but another useful function, which must not be forgotten, is their attractiveness to greenfly (aphids). Whether hardbill or softbill, the aviary inmates will eat these small insects with much pleasure. Blackfly do not seem to be so popular, although some individual birds, regardless of species, seem to enjoy them. Dessert apples should always be available to seedeaters and insectivores, whole or quartered for the former and minced in the food mixture for the latter.

The use of insects such as mealworms should not be necessary during the winter months as the correct insectivorous mixture ought to be adequate. The livefood should be looked upon more as a bonus. Although it is pleasant to see wintered stock acquire a tameness and confidence through the daily-proffered insect life, be it mealworm, wax-moth larvae or locust, even such tiny species as goldcrests can be kept in excellent health for a number of years during which time they receive no mealworms at all in their diet. In the old days such a thing was unheard of; mealworms were considered essential to keep birds singing.

We have, under licence, taken goldcrests from the nest at the time of hatching, keeping them in a wooden box lined with 2.5 cm (1 in) thick polystyrene, the nestlings being held together in soft paper tissues. The young goldcrests have been reared on wax-moth larvae and softfood, the latter being specially adapted for their needs. There were many additions to this diet and it was varied as they progressed and started to peck at the food for themselves. They were, finally, gradually 'meated off and fed on the softfood alone and were maintained on this staple diet all their lives (almost 5 years), merely taking small insects during the summer months when released into outdoor quarters. This illustration of feeding is used to show that a softfood may be up-graded sufficiently for such a purpose. It is not recommended for general use – these birds were taken for research purposes – but their normal estimated life-span in the wild would be a mere 18 months. During the course of the experiments, they received many additions to their basic diet, the most common being Cheddar cheese, finely grated; fish roe was used quite extensively, as were a number of high-protein invalid foods supplied for the purpose by their various manufacturers.

During the winter months, when birds are frequently housed indoors behind glass, one can make good use of rosehip syrup and a number of other such supplements intended for human consumption. They are easily adapted for bird use and can be given either in a drinking font, mixed in water, or in the daily food intake. Do avoid putting such rather sticky items in water containers in which birds may attempt to bathe, for the plumage becomes soiled and unpleasant for the bird. Many brand names, such as Virol and Marmite, and the baby foods of strained beef extract, spinach and such-like, will be known to the reader; these have all been used at various times with excellent effects and using items of proven nutritional value makes feeding a much safer task. The manufacturer will often provide an analysis of his product and go so far as to advise where necessary on the percentage of the product in the total food intake.

Insect Culture

On discovering that an insectivorous pair of birds have young, many people tend to offer immediate large quantities of mealworms and such-like. If this food is too readily available when the young have just hatched, their appetite, small at this initial stage, is very soon sated and it seems that it is then, when they fail to gape, that the parents have an impulse to eject them from the nest. An offspring which lies still and does not gape, seemingly dead, is removed.

In the wild, birds must hunt food to live and rear young and, of course, the close proximity of food dish and contents in an aviary obviates this necessity; this is one of my reasons for recommending that almost all livefood is cut into pieces and mixed with the daily softfood mixture. There are further methods possible whereby a certain amount of hunting can be induced; some small insects, very simply cultured without any elaborate laboratory conditions, can be tipped into a tray holding about 1–2 cm (about ¾ in) of wholemeal wheat. This depth will hide some of the insects, but it will take a short while only for a bird to become aware of their presence. While the parent has the time-consuming task of hunting, a newly hatched nestling has time to digest the previous feed and be ready to gape for the next. I have used both these methods, and others yet to be described, and my personal preference is still to mix the cut-up livefood with the insectivorous mixture. It matters little how wild my aviaries become, or how much natural insect life is produced in the vegetation of them, the constant search for pieces of mealworm or moth larvae in the softfood at every feed ensures that there is no diminishing intake or development of gradual aversion or rejection, necessitating fresh 'meating off' operations each autumn, with the decrease of wild insect life.

Those able to spend a few minutes each month on insect culturing will have a wide variety of food to augment daily basic intakes. Few of these insects are a complete food in themselves but all can be extremely valuable when used in conjunction with other foods. The main essential is a bird-room cupboard or small room – even a disused wardrobe can be converted – which should be maintained at a steady 25° C (75° F). The foods needed for insect culturing are all easily available and the subjects themselves need very little attention. From choice I check my own supplies almost daily but only because I am interested in the process. The life-cycles vary from one species to the next, but average between 30 and 50 days. The female of some species can lay thousands of eggs, so giving a good yield for a small initial financial outlay. Very little attention to the process is needed. Keeping such an insectary at the recommended temperature in the bird room itself ensures that the room remains warm but not hot, seldom if ever exceeding 10° C (50° F) and thus certainly not too warm for any avian inmates; I have very often utilised this excess warmth, allowing it to escape through vents into the bird room, for the general benefit of the birds and my own comfort during the colder months.

Professional entomologists and pest-research establishments, with fully equipped laboratories at their disposal and experimenting upon the control of insect pests, breed their specimens in large glass containers. These stand in trays of oil so that any escapees do not travel very far. We, however, are dealing with harmless creatures so there is no need to take elaborate precautions; in most cases where birds are kept, such an escape would become a happily anticipated event.

Plastic confectionery jars in use today are ideal for our purpose, although moths, when breeding, need a container from which they may drink, otherwise their short lives will become even further limited and little will be achieved in the way of insect production. There are three well-known methods of providing drinking facilities for the moths and I have used them all quite effectively. A glass test-tube can be placed inside the main container with the moths; this tube should be about 15 cm (6 in) long and full of water with a loose-fitting cork allowing a strip of blotting-paper to be placed into the tube so that the end protrudes above the cork; the water will rise by capillary action, enabling the moths to drink from the end of the strip. The tube of water could hold a shaped piece of sponge rubber instead, once again protruding from the top but filling the tube entrance to prevent the moths drowning. The third method is to make a horizontal slit 2.5 cm (1 in) wide by 2 mm (⅛ in) just below the jar shoulder and to fix a water container to the side of the jar with strong adhesive tape; place a strip of blotting-paper through the slit with the lower part outside and hanging in the water reservoir. I would remind the reader that it is only moths which need to drink and any hole such as the slit just mentioned should be covered with adhesive tape when only larvae are using the bottle. They can be a nuisance if they escape; it happened to me once and they entered a nearby thermostat, pupating inside and welding the points together with their silky-like cocoon so that the heater was permanently set on.

The plastic lids of these confectionery jars are a screw-on type and can have a section cut from them and a piece of fine metal gauze fixed in position with adhesive tape. Very thin flexible wire gauze can be held in position with strong elastic bands. This air vent must be of metal as cloth will be eaten through in a very short while. In some cases I have cut a circle from the lids and, with the handle removed, secured a woven wire tea-strainer over the hole. Unless one provides rolls of corrugated paper or similar material to assist pupation, the larvae will mass at the top and stop the flow of air.

Having prepared a number of jars in this manner (to commence I would suggest between three and six for a normal small collection), the next thing is to obtain the nucleus culture. Do prepare the breeding jars first; I have seen a number of such cultures die off where the reception and food supply had not been amply considered.

The food should be prepared next and it will vary according to which insect is being raised and what it will finally be used for. My own choice for a good all-purpose culture is the American wax-moth; it takes a good-quality food which is subsequently passed on to the birds and it can be harvested at any stage between its minimum and maximum growth. It can be anything from a small caterpillar-like grub to a creature double the diameter of a meal worm and a good 2 or 3 cm

(¾ or 1 in) long with a skin as soft as the green caterpillars of the green tortrix moth or large cabbage-white butterfly from oak or cabbage. It needs merely the gentlest pressure to burst and is certainly superior to the tough skin of a maggot or the shell-like covering of the mealworm. Its culture, along with that of several other useful creatures, is described below. Each culture jar will need approximately 500 gm (1lb) of food.

Ephistia cautella
A common European meal-moth, fast reproducing and requiring a minimum of care, is readily taken by any small insect-eating bird. It is sometimes found breeding in a box used for housing mealworms, or maybe in a seed bunker. It is small with variegated wings and the larvae are white with a brown head; the overall length is in the region of 1 cm (⅜ in) when fully grown. An ideal culturing agent is five parts (by weight) whole wheatmeal, to one of glycerine, and a generous sprinkling of a yeast for bird or animal preparation. Each female, when breeding, should lay anything up to 200 eggs over a 7-day period. The life-cycle is on average 45 days at the recommended temperature and conditions. To assist at the pupating stage, coils of corrugated paper rolled from strips of 2 cm (¾ in) width should be a great help. If two or three of these are put in each jar not many will be lost at this stage.

Anagasta kühnella
This is the Mediterranean flour-moth and very similar in appearance to the above moth. The larvae when fully grown are the size of the green tortrix caterpillar found hanging by a thread from the oak in late spring. In colour they are pale pinkish-white, soft-skinned and always welcome when offered to hardbills, as well as to insectivorous varieties. The feeding is simplicity itself since only whole wheatmeal is given, with a sprinkling of yeast. They should reproduce in more or less the same quantities as the aforementioned variety. Again the provision of the corrugated paper rolls will greatly help when the pupating stage is entered. The life-cycle is approximately 50–55 days.

Achroia grisella
This, the lesser European wax-moth, is native to the British Isles. The larva reaches a maximum of 1.25 cm (½ in) in length. The little extra work involved in the food preparation for these is well worth the effort. Even in the fully grown stage just prior to pupation, the larva is readily taken by goldcrests whenever they have the opportunity. In hand-rearing experiments, the larvae have also produced excellent results. The nourishing foods which make up the culture medium for this moth are

as follows: six parts by bulk of whole wheatmeal, six parts rolled oats, one part glycerine and one part honey. When mixed to a crumbly consistency, add a sprinkling of yeast and mix well into the food. The food medium should end up moist but not of a gluey nature and, even when it has been given rather on the moist side, the larvae have been produced in good quantities.

Galleria melonella
The greater European wax-moth can be treated in an identical manner to the last variety. This one will reach 1.5 cm (5/8 in) in length and be proportionally fuller in girth. Again it produces very soft-skinned larvae and this is vitally important where newly hatched young birds are concerned, whether they are to be reared naturally by the parents or by the keeper. An ideal nucleus culture would be about 200 moths or larvae, but, if using moths, remember to provide drinking facilities.

Achroia melonella
The American wax-moth can be fed on the same food as the two wax-moths previously mentioned, but there are other food mixtures which can be used; these do not accelerate production or increase the size but may possibly improve the general condition of the insect and consequently the breeding strain. In the wild these wax-moths will invade a badly infected beehive where the swarm has been considerably weakened and will eat all the honeycomb; this can only be achieved where bees have contracted some disease and are too weak to defend their hive from predators and it really constitutes nature's way of cleaning up after an infected swarm has died off. The skin is very thin and the larvae can reach over 2.5 cm (1 in) in length and far larger than a mealworm in diameter. Even so, any bird seems well able to take them, the smallest of birds beating them against perch or on the floor; certainly only a few sharp pecks render the larvae hors de combat! Other media such as wheatmeal, rolled oats, brood comb, yeast and honey, or Farex (Glaxo), yeast honey and glycerine have been used satisfactorily.

Carpophilus dimidiatus
The com-sap beetle is only barely 1.5 mm (1/16 in) in length but it produces a larva just about three times this length. Despite this rather small size, birds seem to enjoy them. Where one offers an experimental handful of their food medium holding dozens of the larvae, birds do certainly make a feast of them. I collect the beetles in the normal way by placing a carrot in their container overnight; it is a simple matter the following morning to lift the carrot, best cut in half length-wise, and find hundreds of these

larvae feeding on the vegetable. I used to shake them into a shallow dish, to which I would add two or three drops of cod-liver oil, not only to prevent any escape but to render the insect harmless, since they are capable of eating their way out of a cardboard box. The manner in which I saw redstarts swallow them whole and alive made me wonder whether it might be harmful and now I use only minute quantities of the oil, as very small doses will do no harm. The food for rearing this insect need be no more than rolled oats but I always added a good spoonful of molasses and a little yeast. Because of their diminutive size, their content seems unlikely to be of great nutritional value yet they do provide variety if included in an otherwise rather austere diet.

Tribolium castaneum

This rather distant relative of the mealworm, which can be friend or foe, depending upon its use and method of being fed to one's birds, is the rust-red beetle, which, unlike the mealworm, can be fed in any numbers. This is another insect which I have always treated first with a minute quantity of oil; the oil can be corn oil, olive, sunflower oil or cod-liver oil. As a member of the family Tenebrionidae it is most likely the commonest pest of stored goods, being found in cereals and their products, oilseeds, even bones and, in particular, many Oriental dry foods. It is a lighter colour than the previous beetle and measures approximately 2 mm (1/8 in) in length. It produces a larva over double this length when fully grown and, at this stage, it resembles the mealworm and could quite easily be mistaken for it. As a relative of this insect it has the same distinctive flavour and seems quite a favourite with birds. Wholemeal wheat and rolled oats with a small amount of molasses and yeast will produce large quantities of this insect. As with most small insects which go through a pupating stage, rolls of corrugated paper in the culture jars seem to assist them greatly through this part of their growth.

Tenebrio molitor

The mealworm is valuable as long as it is not used to excess so that the bird's digestion is abused. Given in excess it can cause gout in which the kidneys fail to eliminate certain waste, and uric acid crystallises in certain areas, particularly the outer extremities. Used wisely, however, it is one of our most useful foods for birds. I treat them with a few drops of oil before cutting them into small pieces and mixing with the insectivorous birds' basic food supply. The mealworm may be given alive to the hardbill, although I find the best way is to pinch the head, thus rendering it stationary so that it cannot escape from the food tray or bury itself beneath the seed and so be missed by the bird for which it is intended.

Excess moisture is the main danger to guard against when culturing this insect. If one uses a container made of glass, tin or plastic, which does not breathe as does a wooden box, the urine of this insect will build up and cause a deterioration of the food supply, possibly with fatal results for the culture. Minute insects, resembling a white dust, will form over the surface of the container and breed on the decaying food and, in such an environment, the mealworms will soon die.

A dry, well-aired container, a small amount of carrot – so that there is always a little available in a fresh state and it is never allowed to accumulate or begin rotting, a small quantity of rolled oats and wholemeal, with a little molasses and a good sprinkling of yeast mixed well to a dry crumbly mass are all that is required. Remember that the beetle is able to fly so construct the box, if of wood, with a lid of very fine-mesh wire gauze in a tightly fitting wooden frame or, if using a confectionery jar as previously described for moths, lay it on its side so that a maximum area of surface is obtained. Into such a jar or box, lay a double sheet of canvas sacking and, on top of this, place a heaped tablespoonful of the prepared food and a slice of carrot.

To start a culture, place about 50–100 beetles in the jar and add a label outside showing the date. After 6 weeks the beetles' productive life of egg laying is more or less ended. The beetles can be collected, killed and frozen; pass them through a mincer at a later date and add to the food of redstart-sized species and above, but never to that of the small warblers or goldcrests. The eggs will be left on the canvas and on the base and side of the jar if it is positioned as suggested. Place 500 gm (1 lb) of the prepared food in the jar and stand it on its end or side in a warm site. The newly hatched mealworms will soon show themselves and, as soon as they grow to just over 1 cm ($3/8$ in), place a slice of carrot or apple in with them. Do not overdo this for fear of excess moisture which will develop with the first signs of rotting.

If there are about a dozen jars in use then there should be no further need to purchase this insect, unless a very big stock of softbill birds is being maintained. Start one new jar every month and a steady supply should be available throughout the year.

A word of warning should be given on the old-fashioned method of breeding mealworms by placing them in a tea-chest amid two or three old sacks and large quantities of food. The mealworm is cannibalistic and many at the pupa stage, when most vulnerable, will be partly eaten. If one has such a box containing mealworms at various stages of growth, the population will reach a certain point, then move no further; of each beetle's eggs laid there will be hundreds which never reach maturity. The method recommended above ensures that no beetles are left to prey upon the smaller and earlier stages. All will grow more or less together thus eliminating even further the possibility of any cannibalism.

The simplest method of removing beetles is to remove the carrot for 2 or 3 days, then to place a large slice in the jar late one evening; the following morning almost every one of them will be on the carrot and the removal of this should also take the bulk of the beetles with it. The remainder will have to be removed by hand at that time to be sure of a good harvest.

Fly Maggots

So many varieties of fly exist that if one makes a hole about the size of a pencil in a fresh egg and then stands the egg in a dry warm spot with the hole uppermost, within a very short while a horde of maggots will fill the egg, the result of a fly having laid her eggs in the opening provided. Although not suggested as a method of culturing, this illustrates how simple it is to produce these insects.

It is most important to know what food the maggot has eaten. This larva is not very nutritious when kept for any length of time as it then has to live on its own fat and its food content therefore lessens. It is vital with many maggots commercially bred for angling that they be cleaned thoroughly by being placed in bran for about 5 days after having been removed from the source of their food supply. A maggot raised on fish should be a little more nutritious than one grown on meat. A supply can easily be produced during warmer months by using fresh fish each time. Do not allow the food to build up with left-overs from a previous hatching and rearing. A bucket holding bran or sand, or a larger container such as a barrel may be used. Over this, hang a bag made of curtain netting into which has been placed the fish. Place a lid over the whole bag and the container below, so that no rain may fall on it. Flies will invariably visit such an attraction and, within a few days, there will be hundreds of small maggots. As these grow to the stage where they no longer wish to feed but only to pupate, they will leave the fish and fall into the container below. Normally it is only when the old used fish is moved that an offensive odour will be given off; the bag and the remains of the fish can then all be buried in the garden and a new start made with fresh fish. The gamekeepers of years gone by hung the birds of prey, shot in defence of their pheasants, on the nearest wire fence and the consequent maggots falling from these were welcomed by the birds. One seldom heard of a case of botulism in those days for, with a fresh carcase for each breeding, it was most unlikely ever to occur.

When feeding birds with maggots, there are a few additives which one can use to improve the nutritional content and the general value of this creature: to a cup of maggots add ten drops of cod-liver oil, then sprinkle with calcium lactate, Vionate, Casilan or Complan. The oil must be used sparingly – too much can do more harm than good – though the small quantity mentioned will add a slight film to the maggots' surface and ensure that the other fine powders adhere to them. The calcium is to make certain sufficient reaches the nestlings; Vionate, Casilan and Complan are all excellent in their own fields. Used in this manner, the almost worthless stored maggot becomes rich in nutritional value, the fine powder being passed to the nestlings.

Bee and Wasp Pupae
These are exceptionally good food value although culturing is not possible and an apiarist should be consulted. There are usually a certain number of bee grubs wasted in any apiary establishment and they are well worth the little effort in collecting them, either for immediate use or for freezing. They may be given whole to large birds, cut into pieces for smaller species or minced for the very delicate-billed specimens. Likewise the grubs of wasps are a useful addition to a bird's diet and an enquiry at the local council pest-control office may be helpful. Ask if there is any chance of obtaining the wasp nests which have been removed by the department. Even after extermination of the swarm, the larvae should be untouched by the chemicals used since each is sealed in its own little cell. Check, however, with the pest-control official that the chemical is not a liquid which can contaminate the sealed grubs. Using a pair of tweezers, one can open up the full cells and gently remove each grub. These can be fed immediately or frozen for future use.

Locusts
Large plastic aquaria, quite reasonably priced, are easily available and one of these, or a wooden-framed plastic container, or even an unwanted glass aquarium, can be used for the bulk of the locusts. A lid of perforated zinc should be made in two separate pieces: one will hold the electric bulbs which provide the necessary heat and the other, smaller if possible, will act as the door for feeding, watering and removing the females which have been fertilised. Two bulbs are recommended in case one should fail during use, causing loss of breeding stock. Carbon-filament bulbs give more heat and adequate light for this purpose and it is advised that a qualified electrician should install such equipment as this kind of bulb in particular needs a good heat-resistant holder and cable near the fitting.

A main breeding compartment and separate hatching jars are suggested; the adult locusts are kept in the former where they should receive fresh grass and a small shallow dish of bran daily. A sponge protruding from a small glass bowl of water will supply all drinking needs. Have a confectionery jar ready for the fertilised females when they are ready to be removed from the main culture and place moist sand in it to a depth of about 10 cm (4 in). A small piece of sponge in a dish, such as a cage drinker, with water, and sufficient grass for each day should be supplied; making sure that stale grass does not build up and cover the sand base. The female will insert her abdomen into the sand as far as she can reach, then lay her eggs. She will gradually extract her body until the last eggs are just below the surface of the sand and will then seal the egg 'tube'; before putting her back with the adults give her an opportunity to lay other eggs. A container such as a confectionery jar can take up to five fertilised females, but only if they are placed in it on the same day or over a period of 2 days; all the eggs in one jar will then hatch at the same time. Allow just over the 2 weeks' expected period for hatching. The most suitable breeding temperature is about 26°–27° C (80° F), so if housed in an insectary which is controlled at this temperature, the bulbs fitted to the adults' container need be for lighting purposes only; they should be turned on for roughly 16 hours per day. The moistened sand for the eggs to hatch in is made damp only on the one occasion prior to the fertilised female being placed in the jar. The sand should be made just sufficiently damp so that, when gripped tightly in the hand, it retains its shape when released but will crumble under the least pressure. Newly hatched young locusts may be fed directly to birds but I prefer to see that they eat grass beforehand and, even then, I do not feed them to the stock until they are at least 24 hours old. Almost all birds can cope with this insect at this stage, except perhaps treecreepers and goldcrests. For these smaller birds, young locusts can be minced and added to their mixture.

Caterpillars

Caterpillars can be propagated in vast numbers from a nucleus culture which can either be purchased from a laboratory supplier of live specimens or simply procured from the wild. Large or small insectaries can be used for the purpose but this method can be time consuming, mainly because fresh food must be gathered daily and, if the correct leaves are not available at all times, the caterpillars will be lost.

There is a quite simple and interesting method of obtaining caterpillars for breeding and harvesting which is well worth the little effort involved. The old professional butterfly catchers and breeders used a system of tree 'sugaring' when moths first appear in May. Firstly, one must purchase a few ingredients: 500 gm (1 lb) each of Demerara sugar and treacle, plus a small bottle of Guinness. These three items should be boiled together, with constant stirring, until the mixture reaches the consistency of thick custard. An optional extra at this stage is a tot of rum; old collectors added this for the aroma it imparted and it does seem to work. The finished product is termed 'sugar'.

The best source of moths is an orchard, spinney or woodland. Select about four trees and, using a clean brush, paint the bark with a band of 'sugar', 56 cm (22 in) deep, right around the bole at eye level. The same trees should be treated each evening, since the aroma develops and becomes more powerful during the day. This sugaring should be kept up until the required number of moths, of the species which it is intended to propagate, have been taken. Refer to an identification guide on moths rather than risk destroying an endangered species. Any surplus of common moths can be fed directly to softbills needing such livefood.

There are some 25,000 species and subspecies of moths and up to 70 species of butterflies in the British Isles; the choice of species, however, is really governed by the availability of their food supply. When given fresh leaves of the appropriate kind, some moths will produce up to four generations per annum, whilst other species will have just one batch of offspring per year.

The two most suitable moths for breeding are the poplar hawk moth, *Smerinthus (Amorphus) populi*, and the early thorn moth, *Selenia dentaria*. When fully grown the caterpillars of both reach about 4–5 cm (1½–2 in) long and up to pencil thickness; these, of course, can be harvested at any stage of their growth. Each moth can lay up to several thousand eggs.

One needs at least four female moths to each male, those with the larger abdomens being the females.

If the trees are 'sugared' as advised, the moths, being nocturnal, will be attracted and very soon caught in adequate numbers for breeding. After being removed, very gently from the sugar each morning, the moths should be placed either in a temporary container, such as a plastic confectionery jar with suitable fresh leaf food, or directly into a butterfly breeding 'sleeve'. The poplar hawk moth will eat leaves of poplar, sallow willow or willow, whilst the early thorn will take alder, hawthorn, sallow willow or sloe (blackthorn).

The breeding sleeve is a long muslin tube, about 112 cm (44 in) in length and 42 cm (16½ in) in diameter. To use the sleeve, secure it to a selected branch of the appropriate bush or tree by placing the branch in the sleeve and tying tightly around the sleeve top, with the lower end falling just clear of the branch end (see Fig. 2). Place the moths in the sleeve and secure the loose end of the sleeve with string. Within a short while of putting the male and female moths together, mating will take place. The females will commence laying eggs shortly afterwards; these are bright green and are laid singly on the underleaf, where they shortly hatch and start growing.

Harvesting can be carried out whenever the bird keeper considers the caterpillars have reached a suitable size for his birds. The breeding sleeve should be removed and the by-now defoliated branch should be cut and placed in the aviary where the caterpillars are needed. Alternatively, the caterpillars can be shaken on to a large plastic sheet for sharing among a number of aviaries. If caterpillars are bred in an insectary, it is imperative that fresh new branches, bearing sufficient leaves, are added daily in a water container; both water and leaves must be maintained in a fresh condition at all times. When harvesting caterpillars, always set a number to one side on each occasion, this is to allow them to complete their metamorphosis to the imago stage and thus form the basis of future cultures during the same year, or to hibernate, with continued fresh leaves, until the autumn, when wild specimens have disappeared. Both moths and chrysalids can be over-wintered in a dry cool frost-free area, thus enabling propagation to be started early the following spring. It also obviates the need for further sugaring.

Ant Larvae
These so-called 'eggs', particularly those of the wood ant, are a nutritious item too good to be missed if local to one's home. I have moved whole nests into aviaries, the best time being the month of May, and watched the ants soon re-build with the material I had packed with them in large plastic bags; they started breeding shortly afterwards. Ants can be fed by filling a jam-jar with wholewheat meal on to which should be poured hot water containing melted crystallised honey and a little yeast. Allow this to set and cool before placing near to their nest. They will soon make determined inroads into the food. Lay the jar on one side to prevent rain destroying the food supply. A long cane can be used to reach the top centre of the nest when the 'eggs' are available. If the top layer is removed at this time and the 'eggs' spread, birds will soon arrive on the scene for their share of this food. On the other hand this excellent addition to the basic diet can be collected at a wild site. Lay a sheet of canvas in a sunny spot on the ground at least 6 m (20 ft) from the nest and fold over the edges to form a flap of about 10 cm (4 in) all round. Place a few small sticks or dead leaves under the edges of this flap and then gently remove the top of the nest where the largest 'eggs' are stored. Shovel a

heap into the centre of the canvas sheet and the ants collected with the 'eggs' will speedily remove them to the shade of the flap all around the outer edges. It is a simple matter to brush them gently into a container from here.

Fruit Flies
These are simple to breed; one needs only one or two squashed bananas in a jam-jar. Keep about six of these jars active and they can quickly be placed one after the other in an aviary or indoor flight. Another method of producing this fly is to use a bottle of Guinness or milk stout. Fix a drip feed in lieu of the stopper, secure the bottle upside down and site it just above a dish containing a sponge which fills it completely. The flies will produce eggs and the minute maggots will pupate in the drier crevices of the sponge, producing an insect with reasonably good food content.

Insect Collection

A butterfly-catcher's net swept over the uppermost grass-heads and flowers of a field, the contents being removed to a plastic sack or large bag, will in a very short while provide a host of insect life. Bees will soon escape if the open net is held upwards for a moment, so that they may carry on their work undisturbed. A plastic collection jar or bag should be taken on any trip to the country lanes or fields. Even a stroll in the garden can be rewarding for the birds, as a glance under rockery stones and in borders often produces a fair amount of woodlice. Leaves harbouring aphids, and heathers shaken over a plastic sheet or into a bag can usually yield dozens of small spiders and other creatures suitable for any nestlings in the rearing stage.

Where permitted one can use moth traps at night and gather a multitude of insects as well as moths. However, do keep a watchful eye on the species of moth taken. Today many species are so scarce that a little care in the conservation of these is the duty of everyone.

Worm Culture

Whiteworms
These can be cultured quite easily but I find few birds prefer them to the insects mentioned above. Those who do wish to breed them should firstly prepare a breeding container; plastic seed trays are ideal for this purpose. Obtain some peat, sufficient for each tray intended for use; sterilise this

in the oven, then fill the trays and keep them moist, taking care that no other insects can enter. Have a sheet of glass for each tray, just large enough to cover it and overlap the edges. Now, to obtain a nucleus culture, mix half a cupful of wholewheat flour into a paste with water, find a damp, sunless spot in the garden, make a slight depression in the ground and pour the contents of the cup into it; finally place a piece of slate over the top. Add amounts of fresh paste daily. Many other worms will eat it but after a few days small hair-like whiteworms will appear. Keep feeding these until they increase to form a small knot of creatures 3 or 4 cm (1 or 1½ in) across; at this stage gather them and place them in a small tin containing wet moss. Leave them like this for 2 days then very gently, with a spatula, remove them to the prepared tray. These trays must be kept in total darkness. The worms should be fed daily with only sufficient (a teaspoonful at first) for each day so that the paste does not go sour or rancid; gradually the mass of worms will multiply and, once established, will breed quite freely and in large quantities. If the

Fig. 2 Butterfly breeding sleeve in which caterpillars are produced, from a nucleus culture of moths, on natural food.

worms tend to 'ball-up' when collected for the birds, harvest them in the evening, place them in damp moss in the dark and they will be widely dispersed by the morning on which they are required; the moss can be shared between the birds and they will take them as required.

Brandlings
This is a reddish-coloured worm which reaches about 4 cm (1½ in) in length and can be propagated in a similar way to the far smaller whiteworm, but in larger containers. Use a cast-concrete pit, measuring in the region of 2 m (6½ ft) by 1 m (3¼ ft) with a depth of at least 28 cm (12 in). A fine wire or plastic mesh outlet must be incorporated into the base to ensure the escape of excess rainwater but it must be fine enough to prevent the small worms doing likewise. Fill the pit to within a few centimetres of the top with loam, leaf mould, peat or even clean vegetable compost, then add the nucleus culture of brandlings. These can be obtained from any specialist angling supplier. Feed them on a mixture similar to that for whiteworms (above), giving half a cup to begin with. Then increase or lessen the amount so that all is consumed in one day; a fixed amount can soon be estimated fairly accurately in this manner. Never leave sour food for the worms. The pit should be covered at all times with a canvas sheet or plastic woven sacking. This serves two purposes: it maintains the soil in a damp condition and prevents the local thrush family from depleting the harvest. One or two roofing slates laid upon the soil just beneath the canvas will encourage worms to congregate there, ready for collection, and will also give some indication of the numbers of worms being propagated. Allow a short period for mating and for eggs to hatch; pairs of worms will be seen coupling on the soil beneath the slates, making their breeding activities obvious. During any exceedingly dry or hot periods, moisten the sacking and soil with a bucketful of clean water, but, apart from this, do not disturb the compost too much, other than to collect the worms.

Other Foods

One should be careful when considering unconventional forms of food. There was a time many years ago when I wondered if one could breed the beetle from which cochineal is prepared, a small insect reared on cactus in Mexico. I had visions that this might enable certain birds to retain colour which is frequently lost when in confinement. Later in life, when visiting Mexico, I admired the many flowering cacti I saw but left the insects to their own devices after I had investigated the cacti's skin texture. Spiders can be collected but not bred in vast quantities. Crickets can be bred in a dustbin containing pages of newspaper rolled into balls, and fed on rolled oats and wheatmeal at about 27° C (80° F).

Small fish fry can sometimes be collected by the bucketful and can be stored in the freezer or fed minced in softfoods to birds of redstart size upwards. Provide a few each day, alive in a shallow dish, so that parent birds may help themselves as they wish. They will then be seen to hammer the fish well before offering it to their young.

4

Hand-Rearing Young Birds

THIS IS A METHOD OF REARING YOUNG BIRDS to the stage where they can ably feed themselves with no assistance other than the daily provision of food, water and other necessities. For most people this can be a problem as working hours, sleep, hours of light and leisure time rarely coincide with the needs of a young, completely dependent bird.

Warmth is as vital as food and we must be sure that, during the night when temperatures drop to their minimum, the young birds are kept warm. The use of a nest is far from sufficient and many young birds die solely through their keeper's ignorance of this fact. A box, either of wood or cardboard, will be required; line this to a thickness of about 2.5 cm (1 in) with polystyrene, which can be cut from ceiling tiles or obtained from do-it-yourself stores or builders' merchants in sheet form of various thicknesses. The box, including the base, should be lined fully. One piece of polystyrene is needed for a lid and this should have a hole in it about the size of a pencil. The young bird will need to lie in soft tissues, which must be replaced frequently to keep the nestling dry. The excrement sacs, too, should be removed after each feed.

As would be expected, the diets of wild insectivorous and seedeating species vary, although not to any great extent during their first few days as both then take animal protein in the form of insects, and a certain amount of vegetable matter. Gradually, however, the seedeating varieties increase their consumption of farinaceous foods and vegetable matter, whereas the insectivorous birds continue on their all-insect diet with the only vegetable content being that of the undigested food of the many insects fed to them.

The ideal age to start rearing a young bird, if there is a choice, is 5 or 6 days old. Only too often the time is not chosen but thrust upon us when an accident in the wild causes some kindly soul to bring a bare gaping creature to the door with some such comment as 'we found this, can you do anything to help'. Perhaps in our own aviary we may find a youngster on the floor, apparently dead. Where this happens it can be a case of a hen flying straight from the nest and dragging an offspring with her. Very often breathing on the youngster, which should be held in cupped hands in the warm for a few minutes, will revive it sufficiently for it to commence gaping again. Whatever the cause, whether handed to us or taken specifically under licence for one purpose or another, the task of rearing it is before us.

We invariably use a specially prepared Avi-Vite hand-rearing food, although this is no longer available commercially. Full details have been published elsewhere on the successful rearing from the egg of many species, including the goldcrest (*Regulus regulus*). A suitable brooder to maintain natural body heat, particularly at night, becomes as vital as food at this stage.

For the seedeaters I have used insectivorous mixtures made into a paste with milk, to which I have added a pinch of grated cuttlefish or calcium lactate, some cut-up mealworms, chopped dandelion, comfrey or a little brassica. Never mix a great deal at a time: use two teaspoonsful of basic food, with a dozen mealworms and about half the equivalent of vegetable matter as softfood. Add milk to bring the mixture to a cream if the bird is very young, or crumbly-wet if 5 days or over. This can be fed to the bird either by a spatula, shaped at the end, or by a syringe; I have used one originally intended for icing cakes, and adapted the nozzle by enlarging the small round hole. The spatula, although taking a little more time, is probably the best for the inexperienced as the food must be placed as far down the throat as possible whilst the bird is gaping. A syringe can tend to choke the subject in the hands of the inexperienced. If one is used, however, always point the nozzle well down the throat on the bird's right-hand side, towards the crop; after a short while one can fill the crops of a dozen youngsters in almost that many seconds.

Try using Farlene (Glaxo), a baby food which has been extensively used for young birds with good results. Add to it the above recommended extra ingredients and the bird should thrive if fed frequently enough. It is always far better to feed little and often, but always endeavour to keep the crop packed full at all times from early morning until one retires for the night. Then the box containing the young bird, or birds, should be placed near some form of heating; an airing cupboard will suffice if the door is left ajar for fresh air to reach the box's air vent.

When the young hardbill reaches the stage when it greedily makes a grab for the food being offered, provide a shallow dish holding some of the rearing food and sprinkle a little pirella seed on top of it; add a further shallow container holding some soaked seed which should include red rape, teazle and for finches other than bullfinches, chaffinches and bramblings, a small amount of hemp. If the hand-reared specimen is one of these finches substitute a few cut-up mealworms on top of the soaked seed for the hemp.

When the subject has been eating this food regularly, in addition to the food from the spatula or syringe, for about a week, one should then offer the seed mixture normally offered to the stock, but always in a soaked state. Keep the bird on this and the rearing food until it has fully moulted. As soon as the bird is observed to be eating properly from its dishes, offer leaves of dandelion, comfrey or any of the brassicas, not too much at a time but enough to last the day without souring; with the completion of the moult the bird can be run into an outdoor aviary after first having been introduced, for a day, by hanging the cage just inside the enclosure door.

The hand-rearing of insectivorous species is no more difficult. Start off with a small amount of good-quality high-protein softfood, or Farlene (Glaxo), as a base. To this quantity of, say, two teaspoonsful add a dozen cut-up mealworms, hard-boiled yolk of egg, a pinch of grated cuttlefish powder or calcium lactate, and about a third of the amount of the basic softfood and of finely chopped greenfood such as dandelion, comfrey or spinach. Then add milk and, before stirring, add one drop of Abidec. Mix the whole to a cream and always feed softbills with a spatula.

At the age of 6 days one can start giving, well mixed in the softfood, some finely grated Cheddar cheese. If a grater is not making it fine enough, pushing small pieces through a metal tea strainer with the thumb will render it suitable for mixing in the food; start at about 25 per cent of the basic mixture and increase gradually to form a third of the whole food within a week of starting.

As the bird progresses so too can the diet gradually change. In addition to the mealworms mixed into the food, offer from tweezers a few wax-moth larvae between meals of softfood; these should have had their skins burst or heads squeezed with the tweezers before the birds eat them. The time will come when the young bird will peck at the food as it is offered on the spatula and, from this moment, it should be caged; it will soon begin to take food from a shallow dish. I like to offer a container rather larger than a saucer, sprinkled with the food in a little more crumbly texture and with some grated cheese and a few mealworms cut into pieces. Then, in a small area in the centre, I place about 10 or 12 maggots and cover them with a clear upturned glass; I keep an old broken wine glass for this purpose with the broken stem ground down. The bird's attention will be constantly drawn to the movements and, as it pecks at the insects inside the glass, the bill will slip and come into contact with the food; this encouragement to feed is also used in 'meating off'. The moth larvae can be increased to ten or more a day and the mealworms should be correspondingly lessened to five per day. The required calcium will now be provided by the softfood or the Farlene although the latter, if it has been used at all, should have been discontinued in favour of the softfood, intended for general use, at the time of the bird's being caged; at that time the bird will still be taking the rearing food from the spatula as well as commencing to peck at food.

As soon as the moult is complete in these birds, endeavour to give them a few weeks outdoors in a smallish aviary where one can keep an eye on their general health and behaviour. The winter quarters will vary from species to species but the majority can be allowed outdoors on good weather days, and some all the time, just so long as they can re-enter the warmer atmosphere of an indoor shelter. Bathing facilities should be available to them from the time they begin to eat for themselves. As their self-supporting age is reached they can be given Abidec in their drinking water twice a week until the end of the moult. Even whilst hand-rearing is in progress there is much more which can be added to their diet, but for simplicity I have kept to an easy-to-prepare mixture for rearing and general feeding. Consulting the entry for the particular species in Part Two will give more comprehensive details of suitable foods.

5
'Meating Off'

WHEN ANY BIRD IS TAKEN, under licence, from the wild, some knowledge is necessary to prevent dietary shock which can be fatal.

One can usually house seedeaters individually, or with a tame inoffensive bird in a box-type cage. To the drinking water add a few drops of a vitamin food supplement. The seed mixture should be as varied as possible and soaked to the point of sprouting. After a week or so on this diet, together with wild foods, the bird can be given a mixture of dry and soaked seed on a 50/50 basis. The only time adult hardbills should be collected from the wild is in the early autumn or winter time. A few mealworms and maggots in a shallow dish will often be taken but it depends upon the species; wild foods of the bird's choice should always be available at least for the first month under controlled conditions.

The insectivorous bird requires a little patience and one must be constantly alert to the situation. Watch what is happening, not by standing in front of the new specimen but each time one is in the vicinity of the cage by glancing quickly at the food dish.

For those who are prepared to make a special effort, I would advise the construction of a 'meating-off' cage, which ensures the safety of the subject and helps immensely in weaning freshly taken insectivorous specimens on to a home-prepared diet. Having used a cage such as that described below for many years, I can vouch that it adequately fulfils its purpose. Firstly make a cage base in the form of a box about 8 cm (3¼ in) deep. Its area should be roughly that of a canary double-breeding cage. This base should be constructed with two drawer trays so that the finished cage is easy to clean. Purchase some softwood approximately 1.5 cm^2 (½ in^2) in section and from this construct a frame (Fig. 3) to fit the upper rim of the cage-base sides; when glued and panel-pinned together, this frame should be secured to the base with about six small flat brass hook-and-eye catches, two either side and one at each end. This frame should incorporate two doors: one for feeding and watering, the other for removal of the bird without handling or undue chasing or

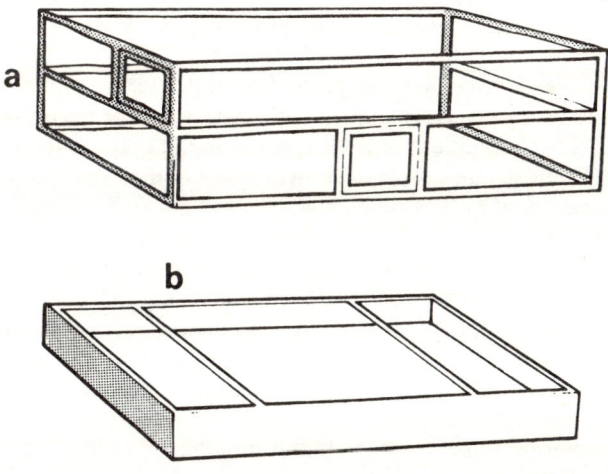

Fig. 3 'Meating-off' cage construction details: (a) frame; (b) base and perches.

shock. When structurally complete, cover the entire frame and doors with a very finely woven nylon material, almost as closely woven as linen. Secure this by applying glue to the frame one side at a time, placing the material in position and pinning it taut on the frame until the glue is completely dry.

This 'meating-off' cage can be used for even the most difficult species. Adult goldcrests can sometimes prove troublesome but, in a suitable cage, even fruit flies cannot escape. Young locusts, blow-flies and moth larvae will remain nearby, tempting any inmate until they are eaten. Give drinking water containing Abidec (Parke, Davis). A shallow container a little larger than a saucer should be covered with a thin layer of a high protein softfood, including about 50 per cent steamed and minced ox heart and liver and plenty of grated cheese. A little extra cheese should be sprinkled on top and upon this should be cut several mealworms. Leave the centre of the dish bare and, in the space, place half-a-dozen mealworms and about a dozen very lively maggots; cover them quickly with an upturned clear glass as described on page 42 so that the birds will peck at the glass and, as their bills move downwards, come into contact with the foods.

There is no excuse for losing a bird taken from the wild: even a box-type cage with a piece of butter muslin, or even hessian or linen, hung over the front will give the bird that required security it needs in the first day or two. Ensure that the material allows adequate light to reach the bird. After a few days this cloth can gradually be raised.

The time taken to 'meat off' a bird varies with the individual as well as the species; I have seen wagtails and nuthatches eating within moments of being placed in a cage with food. I do not like to see any body-weight loss at all and, if feeding is carried out correctly, any bird should be eating reasonably well within 2 hours. Correct feeding involves renewal of the cut-up mealworms and a supply of more cheese four or five times daily. A bird should be treated like this for at least 2 weeks, by which time it ought to be eating everything that is offered. Always ensure that the softfood and cheese are being eaten in quantities denoting a healthy appetite before reducing the number of mealworms, and always check the birds' advised diet before catering for them; some may need fruit, others sunflower seed, berries or a particular variety of livefood.

Finally, mention may be made that nuthatches should not be put in a cloth 'meating-off' cage: they will ruin it within an hour. Use a box cage with lengths of bark secured to the inner sides or propped up with the base very near the food dish.

6

Pairing, Nesting and Nest Sites

GENERAL INFORMATION on this subject may prove a useful addition to the specific comments on each species in Part Two. Some pairs of birds can be placed together and no sign of bickering will ever be noticed but, I would hasten to add, a further pair of the same genus can be freed in another aviary and murder be committed within minutes. The very first thing to realise is that one cannot attribute to birds the human process of thought; because some action is logical to us, we tend to imagine a bird will think likewise. It may well be so in some cases, but we must respect that, if a pair we had thought ideally suited does not wish to contemplate family-raising duties at a particular time, it is better to treat them to a gradual courtship than risk one bird mauling another.

Many birds benefit from being housed in an aviary adjoining that of their prospective mate. At first they may completely ignore the bird on the other side of the wire, or even attempt to attack or threaten it at every opportunity. However, over a period, with the advance of breeding condition, there should be a noticeable change in both birds. With longer days, more sun and a richer diet, the opponent for each crumb of food, the antagonist for a favourite perch, the challenger to the territory, slowly becomes more attractive and nature will take its course in time. What we must do is watch for the perfect moment to open a dividing door.

First study the wildlings. Does the male choose the site of nesting operations? Is it he who first arrives from migration and stakes out a territory? What is the 'key' that unlocks a mating reaction and acceptance? There is always one and, when you know your birds, you will be able to breed them in aviaries. The problem in breeding is not pairing but providing the right substitute food for newly hatched nestlings for at least the first 5 days.

The moment to pair could be when a male enters a nest site and displays to the female, and she attempts to follow him, despite the wire division between them. If she remains uninterested at the far end of her aviary, wait a little longer; nature does. In the wild the male would probably chase her from the territory if she showed this lack of response, then await another female's arrival, attracting and calling her with his song. In an aviary he cannot chase her away and we must here bend natural law a little as we can rarely provide another female in just the right condition. A few more mealworms daily, the concentrated vitamin we should supply to any bird lacking in condition, the male's song, all will play their part in this little drama.

One example does not prove a general rule, but robins can be wretches for fighting between themselves except during the breeding season when a pair will seem the most devoted of couples. We hand-reared three, liberating the odd one when certain we had selected a pair; they were housed in adjoining enclosures and, in this instance, we placed a nest box in the hen's aviary. Originally they had fought at every opportunity and, indeed, I had at one time to separate them forcibly by hand when they were clutched together and ignoring my presence. Gradually, with the male's constant song, the female began to follow his movements, although still in her own aviary. The arrival of a wild pair in my garden hastened things a little as the males spent much time fighting through the outer wire; perhaps this action, so natural in the wild, prompted a reaction from the hen and she was eventually seen entering the hollow-log nest site in her aviary. The male was doing his best to reach her at these times and it was then that a door was opened in the dividing partition; from that moment it was all plain sailing. We eventually had four clutches of eggs and far more young robins around the place than I ever intended. I would mention here that each lot of young had to be removed because of the male's attacks upon them once they reached a self-supporting state. These same young were finally released to the wild when they had been observed taking enough wild insect life to be self-supporting. We liberated them on alternate days: in all probability some were chased from the area by local birds but one of the original pair still lives nearby and visits the garden on occasion.

The common redstart can be an example of difficult pairing, and yet I have known someone who bred them in a very large flight cage, certainly something I would not recommend. On the other hand I have never had any difficulty in the pairing of black redstarts, seldom witnessing even the slightest bickering. Nightingales, wheatears and barred warblers have all earned themselves a bad name in this respect and yet I have known pairs of each which were very peaceful, gentle and devoted in appearance.

The important point is timing. With characters as diverse as those of human beings, birds cannot be predictable in their actions; they must reach their peak of condition to be paired without risk. No matter how long one keeps and breeds birds, there is not one man alive who 'knows it all'; one just cannot give precise instructions and say, 'do things this way and success will be yours'. We have all come across the odd person who believes they have all the answers, but if it were all so easy we might as well keep white mice. The challenge to learn from one's studies is in itself a great attraction, but, were the truth admitted, the longer one keeps and studies birds, the more one is made aware of one's limitations.

All birds require daily bathing facilities; the importance cannot be over-stressed. It is essential for general fitness and for firmness, colouring and well-being of the plumage, but it also contributes to breeding condition and one must never ignore anything which assists the birds to reach the peak of condition.

When pairing birds for the breeding season it must be realised that the adults have to be 100 per cent fit to obtain young of an equal or better standard, or even to obtain full eggs. For this reason, the food supply prior to the beginning of the season should be of high nutritional value. The hens in particular should have sufficient flying space available to obviate any undue fatness, subsequent egg binding and even failure to lay. The insectivorous species should have extras in the way of live insects, though not too many mealworms and these should be cut in half, or at least killed, at all times. The seedeaters too must now receive a few mealworms. At first, cut these up on top of their seed supply and, as soon as they are seen eating them – they can have a daily supply, – just pinch the heads of the insects before giving them. In the case of hardbills this is not a safety precaution but merely to prevent the mealworms escaping before they are eaten. If the seedeaters have been receiving sweet apple, plenty of dandelion, chickweed, or even spinach, comfrey and brassicas of all kinds where wild plants are not available, they should now be coming into full song and chasing their hens. Offer plenty of fine oyster shell and cuttlefish; another useful item, not often used is rock salt, as offered to cattle to lick. It is amazing to see how much time is spent by birds steadily nibbling at a block of salt.

Before the eggs are due to be laid you should make certain that they will be fertile, and that healthy chicks will hatch, by supplying all the essential nutrients well in advance. This should be not merely the month before, but prior to the moult and right through the colder months. Many bird keepers complain of clear eggs, developing chicks dead in shell and similar 'ill-luck', but these problems can be avoided. Do not discard adult birds because of these problems but rather analyse your feeding prior to the nesting season. For any chick to develop from an embryo, all the vital requirements for its growth must be present in the egg to provide it with the strength needed for ultimate hatching and the first 24 hours of the nestling stage.

Sometimes a freshly imported bird or pair will nest and hatch young, only to lose their nestlings after apparently searching everywhere for something that is missing. I have seen this with a few birds from the former USSR and can only conclude that, when first imported, they are unused to British vegetation and tend to search for plants which they are used to. In my own experience, this problem has sorted itself out by the following year and I have always been glad that I did not dispose of the birds for their seeming lack of ability in rearing; the errors were mine not the birds'. When rearing is in progress, the parents often slacken off feeding around mid-day, not stopping fully but resting between journeys; the young, too, tend to nap. The parents resume their activities in earnest at 3 p.m. or so and it is important to make sure that there are sufficient of the right foods for them to give the young prior to the long night ahead, as well as some for an early morning feed next day, whilst their keeper is still asleep.

The pairing of hardbills is often a simpler matter than that of certain insectivorous birds. Normally a pair of hardbills, alone, in a trio or even under a colony system, will accept each other without undue risk of fatalities occurring. It is normal for the early provision of extra rich food supplies – mealworms, dandelions and the growing plant life in their enclosure and even sprouting seed – to take effect on their condition in conjunction with the extra sunlight and longer days. There are, of course, occasions when a pair do seem to dislike each other, but it is rare indeed for harm to result, unless they are large birds, such as the hawfinch, or there are more than one pair of the same species together. Do not place the same trust in softbills.

In the wild state excellent examples of a species may quite possibly choose poor-quality partners, but in aviaries the best two can be paired, ensuring, as near as is possible, good-quality young. If this can be carried out year after year, one can build up a stock of quite impressive quality. At times during my liberty-breeding experiments, when using such specimens, the young have benefited even more from this controlled pairing of ideally matched parents combined with the subsequent natural diet collected by them from field and orchard; these two advantages improved a strain noticeably. Let me here add that liberty breeding is still really in its infancy because the re-taking of the parent birds, as wild specimens, is limited to those with official licences. I do not recommend beginners, even those with a licence, to attempt liberty breeding unless they have unlimited funds to replace possible truants or genuine losses through predators; sadly, the latter can be frequent.

There are no hard-and-fast rules for nesting that one must follow. Once birds have nested, interference should be avoided; one should ensure that no nest is sited where it may become tinder dry, but, apart from these points, it is difficult to aid them as birds can be as perverse where nest sites and nesting are concerned as any animal or human being on their worst day.

Avoidance of too much disturbance to nesting birds is as wise in an aviary as in the wild; certainly there are pairs which constitute the inevitable exception to the rule, particularly where birds have become finger-tame over a period in an enclosure, always accepting the proffered mealworm or other favourite items of food. On the whole, observation from a distance is more satisfactory to bird and man; an ever-alert eye is frequently necessary to be certain that young are being fed correctly and regularly but this is easily discovered without undue and obvious attention.

Nests should not be allowed to dry out too much because moisture in a nest, the warmth of the parent's brooding body, and the slight clinging dampness which occurs, all play their part in successful hatching; so often one sees a hen bath and return to the nest a few minutes later. I well recall an old friend of my father, a real countryman and a breeder of exhibition Yorkshire and Norwich canaries in his day. He used shallow dishes of water in his breeding cages into which were placed the old-type terracotta flowerpots, one to each cage, with an earthenware nest pot in the top: his explanation always seemed reasonable enough at the time: the damp assisted the hatching. The more one thinks on it, the more feasible it seems. From that time onwards I have always attempted to provide nest sites which, although not fully open to the elements, were not completely sheltered either, and I have no complaints to make about developing chicks failing to hatch.

A mouse climbing near a sitting hen at night can easily frighten her from her nest with a resulting loss through chilled eggs. We have looked at the problems of keeping mice out of aviaries in Chapter 2, but we can still do various things to combat this menace inside the enclosure. Rodent poisoning can be effectively practised as long as one makes certain that birds cannot reach the poisoned grain and that mice cannot carry it from its site to a place where birds may partake of it. Traps that take mice alive can be a good investment but the very small ones can escape even from these. With a little thought one can provide many

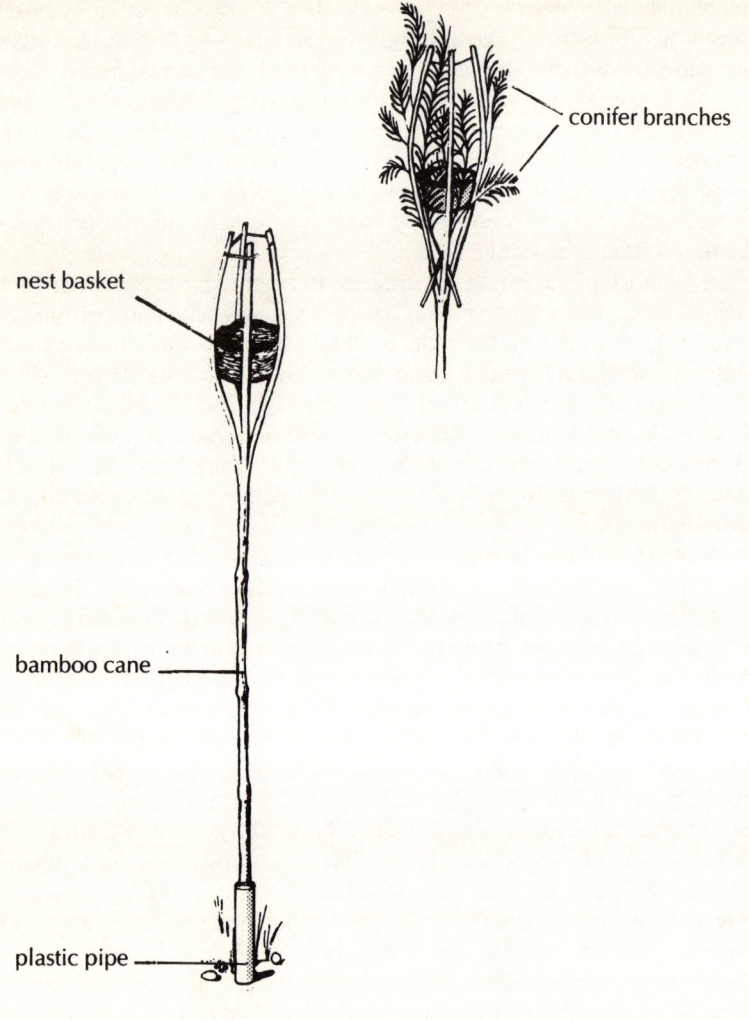

Fig. 4 A finch nest site which should prevent any mice reaching a nest or disturbing a nesting hen in any way. The site can be made simply from a bamboo cane, a nest basket, wire and conifer branches. A short length of plastic pipe pushed into the soil will hold the cane in position and prevent it from rotting.

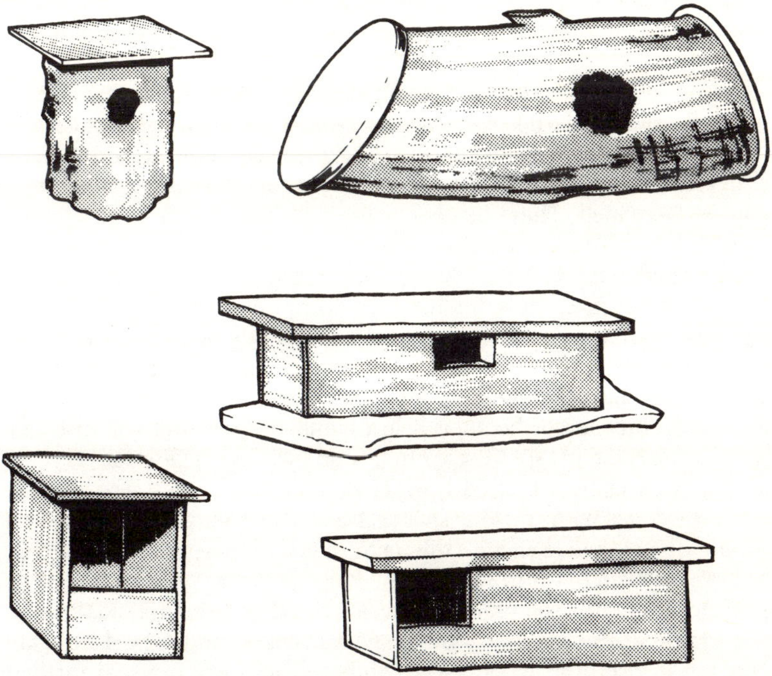

Pairing, Nesting and Nest Sites

Fig. 5. Various log and box sites.

Fig. 6. A timber site with a sheet of roofing felt and lengths of beansticks for the roof.

nesting sites that are almost certainly mouse-proof. Make use of plastic rainwater pipe, ensuring that one end is firmly held in the ground. Block the open end to prevent any inquisitive insect-hunting bird from disappearing into it and erect nest sites on top. Two such posts can be used for the more complicated sites. We all like to give such birds as wheatears a natural site and bird keepers whose aviaries are always free of vermin are very fortunate in being able to do so. However, many are less happily situated and, for those, sites such as those shown in Figs 5, 6, 8 and 9 can be fabricated. The wheatear will readily accept such a site, even though it is not on the ground.

The small plastic cold-water pipe illustrated in Fig. 4 can be used for finches and even many softbill species; using roughly 50 cm (20 in) of this pipe, push it well into the earth of the aviary floor and new sites can be slipped into it as required; the site consists of nothing more than a wicker nest basket, bamboo cane and a few small branches of fir, evergreen, bracken or heather. Select bamboo canes that will just slide into the plastic pipe; split these at one end into four near-enough equal sections to a depth of about 40 cm (16 in) and secure a nest basket about half-way down with the four split sections near enough equally spaced around the basket; then wire the topmost end of these sections together to form an 8 cm (3¼ in) square. The twigs or branches, which serve to provide birds with their required privacy, can then be gently slid through the square at the top, out past the wicker basket and tucked in again beneath it. Use thin twiggy pieces and the canes will not split further. Thin gardening wire can be used to secure the sections a little way apart below the nesting basket, thus holding the twigs in position.

Mice will not usually climb the bamboo canes when a site is constructed in this manner. But beware of positioning it too close to the aviary side, whether timber or wire, as they will certainly negotiate the small space so conveniently provided by climbing to a point above the site and leaping from it to the nest. House mice, which frequent many aviaries, are very adept climbers and leapers, far too frequently being one jump ahead of the bird keeper and thus earning many a hen a bad reputation as a breeding bird, or causing blame to fall on a cat which may even have been chasing the mouse.

One type of nest site that has been experimented with and found highly successful in preventing mice reaching a nest, is illustrated in Fig. 17. It is shown before completion to enable the reader to see the method of construction used.

Nocturnal visits by mice induce night frights in hens thus disturbing them from their eggs or young. These frights can be fairly common, often more so than at first realised. The first indication of this happening is

Pairing, Nesting and Nest Sites

Fig. 7 Nest box, made from willow branches or similar sticks, for finches and large softbills.

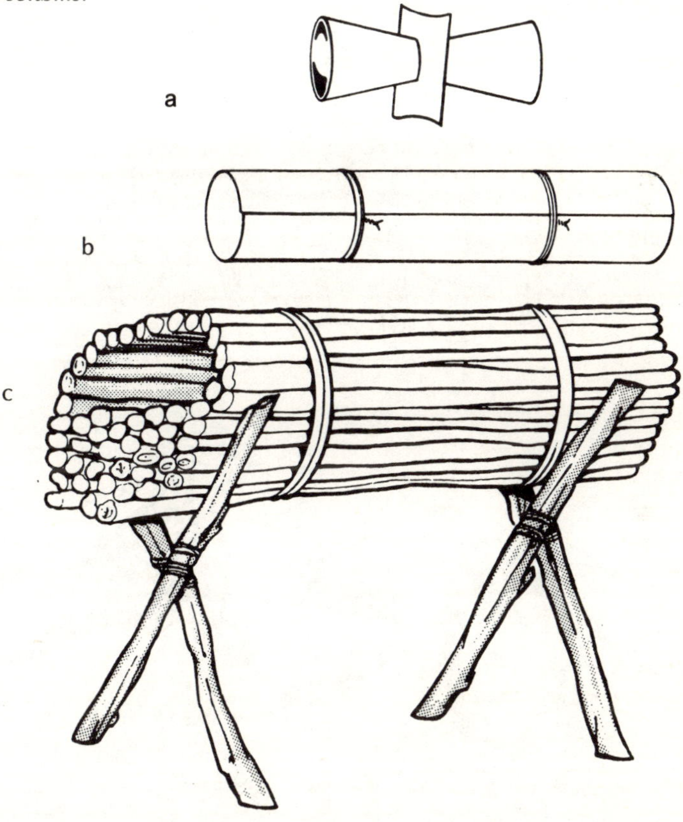

Fig. 8 An 'off-the-ground' site made from two flowerpots: (*a*) position the flowerpots end to end, separated by a piece of roofing felt; (*b*) roll them up in a large piece of roofing felt; (*c*) cover the roll with bean-sticks, held in position with wire, and half close the ends with short pieces of stick. Mount on sticks as shown.

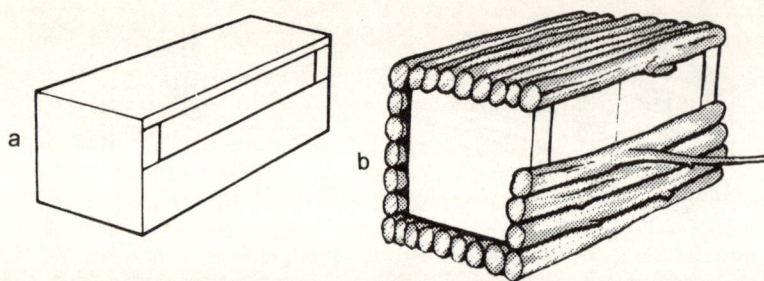

Fig. 9 Nest boxes: (a) half-open-fronted type; (b) hanging nest box made from bean-sticks and two squares of wood covered first in felt.

Fig. 10 To provide a nest site suitable for wheatears, pied wagtails etc: (a) position three or more flowerpots, with the bottom holes blocked, on a mound of gravel which will permit drainage and the planting of grasses etc; (b) carefully place rocks so as to hide all but the small entrance holes.

the presence, in the mornings, of fully-cropped but cold and dead nestlings. Hens are not able to see during the hours of darkness and, if disturbed, are unable to find their way back to their nests.

Using a large plastic or metal lid, at least 28 cm (11 in) in diameter, cover fully a 12 cm (5 in) diameter nest basket or wire-mesh base. Such lids are usually obtainable from drums used in the food industry and should have a deep rim for the best results. With the edge pointing downwards, secure the lid to two 18 cm (7 in) lengths of batten as shown; then take a section of wire mesh bent to form a circle and staple it between the battens at their base. A few lengths of fairly straight twig pushed through the lower holes of the mesh will make a substantial base for a nest.

To fix the nest site to a wire-mesh roof, two pieces of wire should be stapled to the outer edges of the battens so that each protrudes through the lid, where it can be bent to form a hook or 'U'. Provided that a few small conifer twigs are woven into the mesh, so that some privacy is offered to the hen and, most important, that the twigs do not extend too far beyond the downward-pointing lip of the lid, it will be nigh impossible for mice to reach the actual nest, even if they approach it from the roof. However, such sites must be positioned well clear of any means of access from the sides, partitions or walls.

With a little imagination many types of site can be constructed, but, if the pair of birds are really fit, they will frequently make use of something completely surprising. I well remember a pair of tree pipits which had some beautiful sites provided in a large aviary, planted very tastefully and affording them all the seclusion they could wish for, but they chose to nest in the open on a shelf and reared successfully there. Figures 4–17 show a number of possible sites for a fairly wide variety of aviary inmates; the reader should be able to select or amend some designs to suit his birds' requirements.

Do not offer too much in the way of nesting materials to begin with as it invariably becomes soiled while it is being played with. Provide small quantities of various materials, such as teased-out fibrous string, coconut matting, horse hair, dog hair, cow hair, dry grasses of the very fine thin type found with moss in dense undergrowth, small tufts of cotton-wool, moss, birch twigs and heather twigs. Whenever anything is over 8 cm (3¼ in) in length, apart from the twigs, make sure to cut it into shorter lengths to obviate the danger of birds entangling their legs; this can so easily destroy a season's hopes, even if it does not cause the death of the bird.

A quick check in one's record book or card index at this time should show exactly the materials which any particular bird made use of the

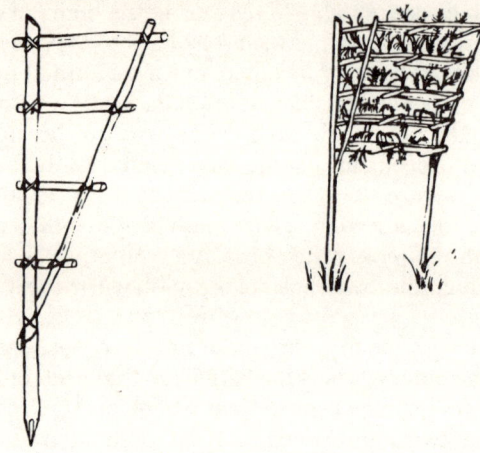

Fig. 11 Artificial hedges of heights to suit a variety of species can be made from bark-covered timbers filled with heathers, conifers, evergreens, etc.

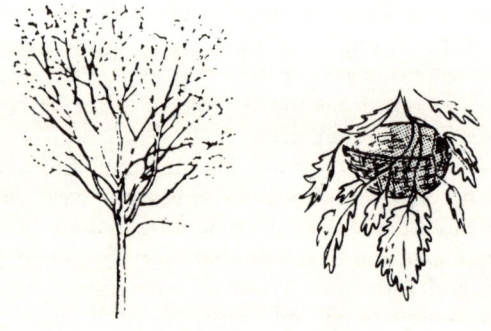

Fig. 12 Even a bare bush or bundle of branches can be covered by beans, sweet peas, etc.

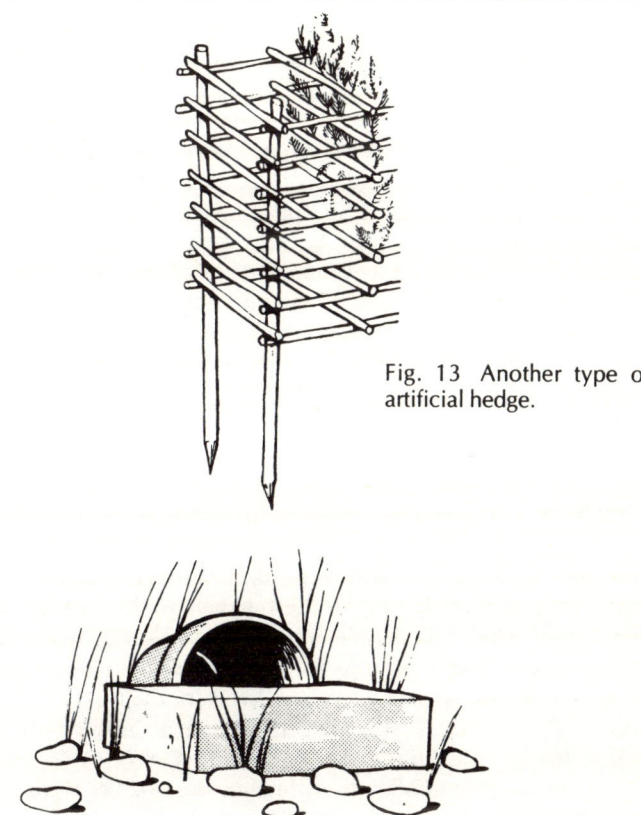

Fig. 13 Another type of artificial hedge.

Fig. 14 Flowerpot and brick site among vegetation.

Fig. 15 A bale of peat covered in seeds, grasses, plants, etc. with nesting holes.

last time that it nested. A lot of time and frustration can be saved when one is fully aware of a bird's choice. Frequently it will be found that pairs, or at least the hen, will wait until after a shower, when basic nesting materials are damp and pliable, before choosing the materials. They prefer the lining materials to be reasonably dry but, where twigs or the coarser grasses, mosses and fibrous materials are used, these are appreciated in a moist condition so that they can more easily be moulded to the required shape. After a quick downpour of rain, or after the aviary floor has been hosed, I have often seen birds readily use nesting materials which had previously been ignored for days.

Make use of wild vegetable down if available, but treat sheep wool from barbed-wire fences with a little caution as I have seen many accidents from its use; either cut it into pieces about 2 cm (1 in) long or leave well alone.

Birds such as the nuthatch will need mud for closing the nest-box entrance and also any cracks and crevices they find in their nest box which allow the light to enter; the thrush family also use mud.

Many species have their own favourite materials: the nightingale likes oak leaves, dead and withered from the previous year, to provide wonderful camouflage for its nest which is situated low in growing vegetation. This will often be amid nettles or low-growing bramble, but often I have had nightingales which nested in other sites; once a flowerpot left on its side was used. Even so, their choice of materials is inherited and oak leaves remain the favourite, mixed with grasses and a little moss.

To long-tailed tits, offer many fine mosses, spiders' webs by the dozen, collected on a thin hooped stick, fine feathers, small pieces of cotton-wool, lichen and small pieces of tissue paper, about postage-stamp size, first rolled in moistened hands. They like to nest in dense honeysuckle, bramble and thick hawthorn, or any similar site in which they can weave the foundation walls of what will be a warm, snug, fully-feather-lined nest.

Often where ground-nesting birds are wary of a ground site, perhaps due to nightly prowling mice, I have found them accept a biscuit tin, fixed on its side to a mouse-proof climbing post. A few nail holes around the top edge, and green turves inside, will sometimes coax a pair into using it. The holes in the roof let in sufficient rainwater to keep the grass inside green and they are provided with ample seclusion. I have seen whinchats take to such a site within an hour of its installation, though a more unlikely whinchat site I can hardly think of. These bogus sites, that can be made absolutely mouse-proof, are a great asset, so do not give up hope when the birds reject orthodox nesting arrangements,

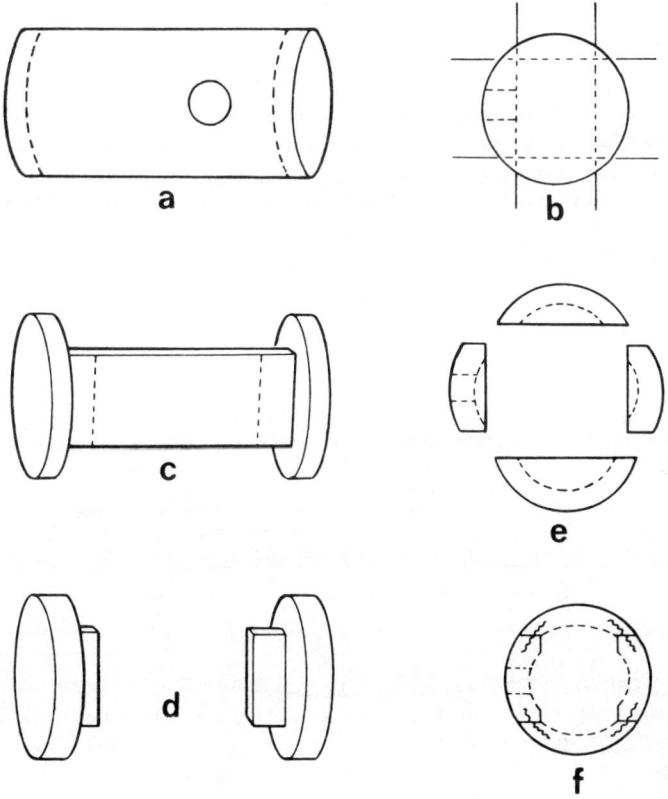

Fig. 16 A simple method of making log nest sites: (a) Take a log of suitable size and make four right-angled (90°) cuts at each end, as shown in (b) the end view. (c) Here, the four sides of the log have been carefully chiseled off and further cuts, as indicated, will give (d) two end pieces; (e) shows an exploded version of the four sides, which should be shaped and joined together again, as in (f), with special timber connectors, obtainable from almost any ironmonger. The two ends can now be replaced and fixed. A square roof can be attached if it is to be placed in very open site in an aviary. The size of the entrance hole will depend entirely upon the species it is intended for; better too large than small is a good maxim.

just the right vegetation, an exact replica of an observed site in the wild etc. They will often adapt; they even have to do so in the wild with man's urbanisation of the countryside.

Observation of behaviour, and record-keeping, at this time can be well worthwhile. Indeed, without accurate data, it all seems rather pointless and there is so much information which still needs to be passed on. Do practise this observation very discreetly so that your presence cannot be objected to by the nest-owners. Sitting outside the enclosure with a pair of binoculars trained on the site will reveal much; one does not have to see into the nest to gather information. Most information can be gained from seeing who comes and goes and what they are carrying to and fro.

Having illustrated suitable nesting sites for finches, buntings and

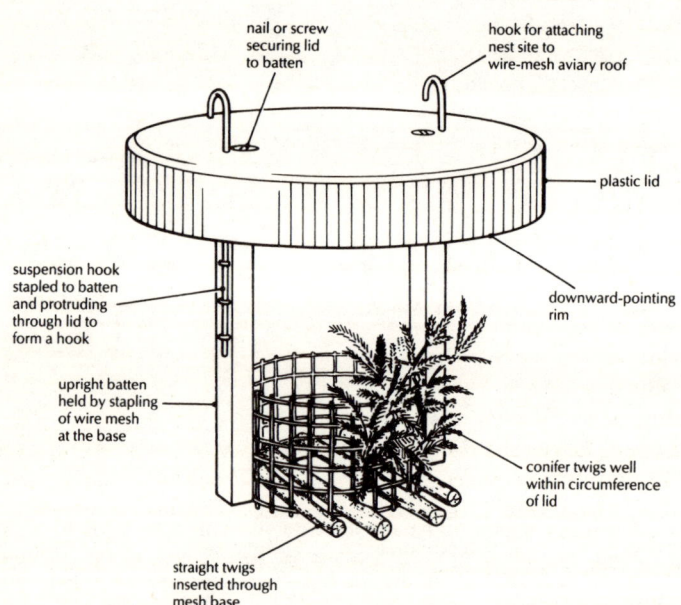

Fig. 17 Mouse-proof nest site.

insectivorous birds, I would merely emphasise the fact that they all have minds of their own, particularly in regard to what meets their immediate requirements when nesting. The presence of nocturnal wandering mice, often unsuspected by the bird keeper, can, for example, cause normal ground-nesting species to seek higher sites or the furthest possible point from vermin. One cannot always blame such untoward behaviour on prowling mice, however, and one must beware of attributing one's own frame of reference to the birds. A mid-height shelf in an aviary must be much the same to a bird as a bank site in the wild, and a half-open-fronted box will provide the same security as the roots of an upturned tree.

Liberty breeding

As the name implies, this method of breeding enables the parent birds to pair, nest, produce eggs and incubate in an aviary but, when the young are a day or so old, they can leave their enclosure and collect a natural food intake for the nestlings. Their keeper maintains the daily ration in variety, providing livefood when necessary, to keep them visiting the feeding area. The system has been used quite effectively with some species. With considerable adult losses at times I have, on a few occasions, been forced to hand-rear nestlings when one or both parents failed to return. Losses can be from cats, predators, poisoning or truancy. I discovered that if I fixed a 30.5cm^2 (1 ft^2) frame over a hole cut in the aviary wire, in the close or immediate vicinity of the nest site, this allowed the adults to enter and depart at will while caring for their chicks. This procedure has only been experimented with in a very rural area, and nearby orchards are ideal. You should allow them to leave only after the young have reached one- to three-days old. Early one morning, open the door and observe them from a distance; the frame, when open, leaves the hole open for free movement back and forth – but this should be covered by a 7.5 cm (3 in) welded mesh to make it cat-proof. Originally I had used the upper corners of aviaries, by removing a small triangular section of metal sheet, which acted as aviary-corner strengtheners, but had found some birds spent far too long looking for an entry. Just prior to the young fledging (one or two days), when you wish to retake the parents, provide a generous supply of mealworms in their enclosure, at a very obvious point, which allows both to enter, and not too far from the nest site. Then close the door with a pull-string. Many species have been bred in this manner, but I also had many losses when first experimenting. On a few occasions young fledged before I expected and were lost, but those which I experimented with were, on the whole, common varieties of which I had adequate pairs.

Part Two

Keeping and breeding details for individual species. The author's successes (and failures) are indicated for each species according to the following key:

† Adult birds hand-reared
* Pair nested
** Eggs laid
*** Young hatched
**** Young reared to maturity
‡ Liberty-bred species

7

Buntings and Finches

WE MAY MAKE CERTAIN GENERALISATIONS about the group of buntings and finches before considering, under separate headings, certain birds for which more detailed advice and slightly different management at certain seasons of the year, or stages of their lives, are required. When pairs are housed individually, one pair to an aviary, life can be very simple. Where mixing is involved, a kind of permutation is sometimes required, certain species being 'high-risk' specimens compared with other more placid types. We must remember to observe all birds as individuals; one can keep a dozen pairs of one species and find that one male or female is as homicidally inclined to its fellow, or fellows (or to members of other species) as the worst criminal in human history – observation will tell us where our trust is being misplaced; if suspicious, one should act quickly and isolate the offender. Having offered that advice, one can go for many years without actually having to heed it, but the odd rogue can do much damage in a short time if not apprehended.

There are some finches which can be colony bred: goldfinches, siskins, redpolls, crossbills and, with a slightly lesser degree of success, greenfinches. The only real disadvantage of this method is the inability of the bird keeper to identify true parentage of the young or maintain any accurate records of stock. It depends entirely for what purpose one is keeping the birds; if it is merely for the aesthetic enjoyment of the aviary inmates' company I suppose colony breeding can be justified; if for behavioural studies, it may be essential. If, however, one was attempting to improve and maintain a true strain of birds by selective pairing, colony breeding is pointless. Bullfinches, chaffinches, bramblings, yellowhammers, hawfinches and most buntings should be housed either as one pair to their own aviary or as one pair to any mixed collection. In some cases they can be very territorially inclined when breeding, sometimes not even waiting for another inmate to cause offence before attacking it; it is obvious that such behaviour is in no way conducive to breeding harmony, even in a large aviary.

Those birds housed together all winter as a mixed collection in a large enough enclosure stand a far better chance of living in peace during the breeding season than a motley crowd placed together at the last moment with little thought given to their well-being or future safety. At times a little belligerence by bullfinches which are selecting a nest site and forming their territory will die a natural death once they have identified that area; quite often things will then remain peaceful throughout the season, the other birds treating the stronger ones with a certain amount of respect. This pecking order is normal and, as long as no physical harm results, can prove most interesting to observe. Similarly, aggressive behaviour can also be expected from the other birds listed as being best paired in their own enclosure. Some may indeed breed amicably without worrying others but the individual bird keeper should keep an ever-watchful eye on birds of this type.

The nesting sites can be various but, on the whole, nest baskets or suitable growing or installed vegetation to provide a choice of natural nesting site and facilities are all that is necessary. The birds requiring different sites are those expected to choose ground, rock crevice, or hole positions for nesting, and comprise merely a few of the buntings. Many of the birds in this category will prefer shelf-type sites, or bush sites, but, unless one is absolutely certain that the aviaries are free of mice, these should not be erected; they could become one of the worst encouragements to vermin and are difficult areas to clear without also putting a stop to breeding operations as a whole. The majority of finches and buntings will accept a basket site or build their own natural nest in a simple site without such elaborate measures being undertaken. However, in very large enclosures, perhaps the walk-through type which may be visited by a number of people, more secluded sites are usual to ensure some degree of security and success.

The diets of this group of birds vary so greatly in the wild that it is almost impossible to give an exhaustive list here. All will survive on the accepted finch mixture as sold commercially but, in the wild, few of these seeds would be taken by many of them. It is a substitute diet, and a fairly well balanced one, but it should be understood that many of the ingredients are not conducive to longevity in the subject. There are so many natural foods that can be obtained or grown to provide variety that one should endeavour to cater for the individual's natural needs a little more closely.

Some birds will over-eat of a certain favourite seed and may become almost addicted to it. Variety, of course, is necessary, but one should not confuse a varied diet with a well-balanced one. Before even a general assessment of the nutritional value of individual seeds or diets can be made, it is first vital to learn about the chemical composition of such foods and the effects of the substances they contain. As a large proportion of the human diet should contain animal, fruit and vegetable nutrients, so too should that of the birds. But birds are no wiser than human beings in accepting the items we know they should have, and the bird keeper's task is to make items such as carrot and celery, or their juices, palatable to the finch or bunting family. Apple is, in fact, easily introduced to the finches. The use of mealworms helps greatly and most birds will accept them after experimentation. The cutting-up of this insect on their seed is usually enough to ensure that it is taken avidly at all times. We can then introduce the necessary moisture to mealworms by feeding them carrot, apple and celery and, if this becomes a general habit, then a certain amount of the required substances will reach the bird in the undigested portion of the mealworms' food. The reference to fruit and vegetable juices is meant as an illustration; they are important, but so are many other foods of proven use; malnutrition in even its mildest form cannot be cured by seed alone and one should make it one's business to learn which foods provide which nutrients and thus prevent certain diseases. Mineral deficiency, for example, causes several diseases which can be easily avoided. A basic knowledge of the constituent elements of the commonly available foodstuffs should be acquired by all serious bird keepers.

Cirl Bunting *(Emberiza cirlus)****

General Description
This cannot be confused with any other bunting: the grey top of the head, black throat and grey-green band across the chest identify it clearly. Both sexes show an olive-green rump. The length is about 16 cm (6¼ in). The female lacks the grey and black about the head and is a paler brown all over.

Habitat
This bird can sometimes be found in similar surroundings to the yellow bunting (page 75). However, whereas the latter will invariably sing from a low bush, the cirl bunting will select a tree, often isolated from others and standing out in fairly open country; they frequent hedgerows and farmland where they gather much of their food from the ground.

Aviary
Almost all the buntings have a tendency to chase the females a lot during

courtship and, for this reason, a lengthy aviary should be the aim of any bird keeper wishing to breed them; I would like to think of an enclosure with a minimum length of 4 m (13 ft) and 2 m (6½ ft) wide. Provide thick, low-growing bushes, not in the corners but standing well out from both the ends and sides of the enclosure, but also give one or two high nesting sites as these do, on occasions, seem to appeal to cirl buntings in aviaries. Where low sites are concerned, allow long grasses to intermingle, growing in and around the bushes which should be about 1 m (3¼ ft) high.

Food
Most of the buntings, despite their love of livefood, are coarse feeders when it comes to grain feeding: plain canary seed, corn, grasses of all types, pinhead groats, the brassicas, millet, screenings and a good-quality insectivorous mixture will all help to keep them in good condition. Most of their natural feeding is done from the ground but in the aviary always endeavour to keep the food on a mouse-proof post site.

Breeding
I have had these birds nest and lay eggs but, on that occasion, something removed them from the nest one night. Shortly after that I passed them on to another bird keeper; one nest was made in a low bush and eggs were laid and the second attempt was in a high site but no eggs were ever found. The nest was made of grass, moss and roots and lined with animal hair, the first one being at a height of 70 cm (28 in) or so and built quite naturally without help.

In the wild these birds rear entirely on insect life and, if success is to occur in an aviary, they should have their own enclosure with adequate supplies of any insects that can be obtained. My adults were extremely fond of young locusts, wax-moth larvae, mealworms and maggots, and all but the last would be suitable for the rearing of young.

Corn Bunting (*Emberiza calandra*)****

General Description
This is the largest of the British buntings and a good male can reach over 17 cm (6¾ in) in length. The hen is always a little smaller and this applies to all the buntings which I have kept or studied in the field. They are not only long birds but very thickset with a tendency towards

plumpness. To describe them as completely brown may sound rather uninteresting but it is not meant to, for with their dark streaking they are most attractive creatures; my corn buntings were brought back from abroad and were never what one could call tame, but I have had hand-reared specimens which were so sure of themselves that they walked over my feet in the aviary.

Habitat

As the name would imply, the corn bunting inhabits arable ground, the stout bill being ideal for the corn that it takes. The penetrating song can be heard from afar across a number of fields, a rather wheezing and jangling one but welcome to the country lover. The flight will be noticed to resemble that of the sparrow.

Aviary

Because the courtship flight involves much high speed chasing, these birds should have an enclosure no smaller than that for the cirl bunting (page 70) with the bushes again about 1 m (3¼ ft) tall and with plenty of long grass. There should also be fairly large clear areas with one or two gravel patches pressed flat to the ground surface.

Food

As well as corn, offer plain canary seed and grass seed of any variety. Some brassicas will be taken, as will a host of low-growing seeds from weeds edging the hedgerows of fields; some vegetable matter will be eaten but insects remain the apparent favourite. A surprising number of hawthorn berries are taken when in the over-ripe stage. Where possible offer screenings.

Breeding

Having hand-reared many of these birds I can vouch for the necessity of much livefood with calcium lactate powder sprinkled on it, but I have reared them to maturity in an aviary although I have not kept many over the years. The pair which nested for me and laid eggs was perhaps not in full breeding condition as two eggs were infertile. The nesting sites require bushes of medium height with plenty of long grass around and growing into them. For any success the birds should be left entirely alone once they are observed to be nesting. The nest itself is typical of the bunting family: very untidy looking, consisting merely of grasses, moss and animal hair lined. Mine was built in April, a little early, which was perhaps the reason for partial infertility in my own pair.

Reed Bunting (*Emberiza schoeniclus*)****

General Description
When in breeding condition the male is very striking with a completely black head, lower chest and belly white and the back a rich brown and heavily striated; in winter the head is browner. Both sexes have a clearly defined moustachial streak, but the female is far more brown and streaky in appearance, as too are the young. These birds are just about 15 cm (6 in) in length.

Habitat
Although most of these birds are to be found in disused gravel-pit areas, well-reeded freshwater streams and marshy areas, I have encountered many in quite open country devoid of anything larger than a narrow stream. In winter months they will move into farmyards and often to suburban gardens in the company of sparrows.

Aviary
The 2 m (6½ ft) by 1 m (3¼ ft) enclosure will suffice for these birds' requirements. Provide some sort of shelter by covering a small section of the rear of the roof, and a few centimetres at the top of each side, just at the furthest part of the aviary from their human visitors: this will prove ample cover for them should they wish to escape from inclement weather or if they are at all shy.

A small pond in the aviary, or a sunken large old-fashioned sink, with heaped-up soil on two sides planted with sedge, water iris and long grasses, will give a very natural setting for these birds; appearance aside, when fully matured, the sedge and grass should also give one or two ideal nesting sites.

Food
Feeding, as with most buntings, is very little trouble and, with adequate supplies of livefood, they invariably come into breeding condition in plenty of time. The basic diet should consist of plain canary seed, brassicas, millet, pinhead groats, many wild seeds and screenings, along with a rich insectivorous food. The usual additives can be well mixed with the softfood and they should also have a daily quota of livefood. This can be varied a great deal throughout winter and late summer but, during the breeding season, they must have wax-moth larvae and mealworms, unless a host of wild insects can be obtained. The maggot proves to be insufficiently small and soft skinned for these birds when newly hatched from the egg.

Breeding

These birds have bred for me on a number of occasions, the young proving quite tame once they have left the nest and acquired some trust in the individual who gives them their daily food. They never seem to build up a strong rapport with the keeper, as do some birds, unless actually hand-reared. If one wishes to install nesting sites, give large clumps of heather and bracken of varying heights, from ground level up to 2 m (6½ ft) or so. Push a clenched fist well down into the dense foliage in a few places to give a cup-like effect here and there and so form possible nest bases. During the greater part of the year these buntings will often forsake most of the seed provided and concentrate upon insects and softfood, but this habit does tend to vary a great deal from one bird to another. The nesting materials to offer are grasses, moss, swept-up dead leaves from the garden and, for lining, almost any animal hair, for example dog, cow or horse, in short lengths.

Lapland Bunting (*Calcarius lapponicus*)**, Ortolan Bunting (*Emberiza hortulana*)** and Rustic Bunting (*E. rustica*)

These buntings may only occasionally become available to the bird keeper and, although I have kept all of them over the years, I would hesitate here to encourage the keeping of them unless for genuine research, or taken under licence for some very good purpose. I would merely add that all can be treated in an identical manner to the reed bunting (page 72), in all respects.

Snow Bunting (*Plectrophenax nivalis*)**

General Description
In breeding plumage the male is a very striking sight, being predominantly very black-and-white; during the winter, however, the colour becomes a varying mixture of browns and white, often seeming to depend upon the age of the subject. Indeed, when watching a flock, I have found that some females and males look very alike until studied closely. The female is slightly slimmer and shorter than the male, being roughly 16 cm (6¼ in) compared to his 18 cm (7 in). She can be sexed easily during the breeding season as she is always a speckled brown but has white underparts and breast; during the winter she, too, acquires more brown in the plumage.

Habitat
Although belonging to barren tundra areas and mountains amid the snow and heather, travelling more inland from the coast in the colder months, the snow bunting is now nesting in Scotland and it is hoped that it will carry on increasing in this reasonably new addition to its breeding range. Not quite as bulky as the corn bunting, but very little shorter, this species will be found gregarious with other buntings although it mixes with few other birds. It can be seen feeding on the ground, taking almost any seed it can find during winter months but reverting almost entirely to insects during the warmer months when they become available.

Aviary
Since so much of the snow bunting's time is spent on the ground, it always seems such a pity when one finds them given an aviary too small to show them off to their best advantage and for them to live as near naturally as possible. The aviary should be a minimum of 4 m (13 ft) in length with a width of at least 2 m (6½ ft). Their normal nesting site will be in rock crevices well shielded from prying eyes or predators. In an aviary one can quite easily prefabricate very near-natural sites with decorative stone from garden centres, or even with bricks and cement. They will appreciate the inclusion of heather in their enclosure, but do leave patches of floor clear for their natural foraging habits. The height of their aviary should never be less than about 2 m (6½ ft) because, although spending much time on the aviary floor, they also fly strongly. Some nest boxes of the half-open-front type can be given at varying heights. Place a few sprigs or small branches of heather around them as I have known very confiding hens to use such box sites.

Food
Despite their taste for a typical bunting seed diet as advised for the other varieties of buntings, these birds do take softfoods during the months when they are taking more insect life than anything else. Their seed mixture should include kibbled groats, plain canary seed, grasses of all kinds, small millets and screenings. Numerous seeding heads of weeds can be gathered from fields and offered to them and, very often, they will wait for the seeds to fall to the aviary ground before collecting them. All forms of insect life are appreciated, even the very small kind of striped snail one finds in nettle beds. I have actually watched them crack these open as if they were seeds; the corn bunting does likewise. During the period when young are being reared, however, they will not want anything but insects for the nestlings, as this is the only way to rear

them successfully. My birds used to take young locusts, mealworms, wax-moth larvae and any other creature offered.

Breeding
The nest-building process of this bird requires far more material than is finally used to be given; so much seems to get wasted and the end product never looks a work of art, being extremely untidy. But, if things progress this far, tidiness matters not a bit. Materials should include dry grasses, moss, dead leaves of all kinds, weed stalks, animal hair – in the case of horse or cow hair cut to 7 cm (2¾ in) lengths, soft feathers and such-like. One of my birds once used a lot of dry chickweed for the base and outer walls. There is a fairly well-established belief among people living in snow bunting habitats that these birds mate for their life-time. I have no proof of this but a well-bonded pair very seldom move far from each other, even when with a number of pairs. I used at one time to have six pairs in a far above average-sized enclosure; the inmates fed alongside each other all the time, weaving in and out in many patterns but always sorted themselves out by retiring together at some time or other in their established pairs.

Yellow Bunting or Yellowhammer (*Emberiza citrinella*)****

General Description
Measuring about 16.5 cm (6½ in) in length, the overall appearance of this bunting, when viewed closely, is of a chestnut-and-yellow bird. In flight, the white outer tail feathers are exceedingly conspicuous; the face, and frequently the whole head, chest and belly, are all a bright, really vivid yellow.

Habitat
Found throughout Europe in moorland, farmland, scrub and almost any area where occasional bushes of a low nature abound, it seems to show a preference for nesting in low sites in well-spaced small hedges; breeding territories are chosen early in the year and are frequented right through until the autumn, when the birds tend to flock.

Aviary
The standard minimum 2 m (6½ ft) by 1 m (3¼ ft) enclosure will suffice, but if a longer aviary can be offered it will prove beneficial later during what can be a hectic courtship display involving much chasing of the hen. A sheltered roofed portion at one or both ends of the enclosure will

provide adequate winter roost and cover during heavy rainfall, snow or frosts. Plant the aviary floor with a couple of gorse bushes and a fairly dense growing hawthorn, some broom and heather, allowing a central area to remain clear. Encourage grass to grow up through any dense-growing shrubbery because this will give the birds a number of natural sites from which to choose. However, do not be too taken aback if they select a high site, even a nest basket, just below the roof.

Food

Generally a coarse feeder like other buntings, it will take much from a soaked pigeon-conditioning base which has had plain canary seed, millets and a little wheat added before soaking; offer also a dry mixture of pinhead oatmeal and ground peanuts, plus a further dish containing insectivorous softfood with the usual additions as for insectivores.

Quite frequently the yellow bunting will accept a softfood made from a human baby food, mixed with milk and with yolk of hard-boiled egg mashed well into it. The addition of a little pirella seed helps them enormously to acquire a taste for this. Once again, if screenings can be obtained, they are well worth the time and labour collecting them from any nearby farm, shortly after harvesting. Livefood – mealworms, earwigs, wax-moth larvae – in quantity is vital when young hatch. They will be grateful for almost anything at such times.

Breeding

This bird will never make a really good job of rearing nestlings if kept short on livefood, so do ensure generous and regular supplies to provide that much needed animal protein. I have watched yellow buntings take hundreds of earwigs which I had previously gathered from rolled-up sheets of newspaper just pushed into a thick-growing Russian vine (*Polygonum*) and then retrieved a day or so later and opened over a plastic bucket. One just cannot forecast accurately where these birds will build their nests. In the wild, nests are often seen very low, almost at ground level, but aviary accommodation, or perhaps nocturnal visits by mice, seem to give them other ideas. Provide dry grasses and plant stems; dried chickweed is often used but in practice many types of trailing weed vegetation will be accepted. Finally, give them plenty of animal hair for nest lining.

Brambling (*Fringilla montifringilla*) ****

General Description
This bird is about 13 cm (5 in) long and is very closely related to the chaffinch, though the general appearance is far darker, the neck and head being marked with black streaks. From the throat downwards the chest is a rich tawny orange. There is a winter and summer plumage, so that, from the breeding season onwards, the head and back of the neck in the male will be almost jet-black. They are simple to sex at all times; the hen resembles the female chaffinch in colouring.

Habitat
Although belonging to northern Europe and Scandinavia where it nests in similar sites to the chaffinch, this bird's breeding range appears to be moving south; it is a frequent winter visitor to the British Isles, arriving in its thousands, but the native habitat is mainly birchwoods. On migration, it tends to flock in open country but often near beeches or hornbeams.

Aviary
As with most species of seedeater, the minimum recommended aviary for breeding should be 2 m (6½ ft) long by the same height and 1 m (3¼ ft) wide; greater dimensions will assist during mating display flights. A roofed portion should be included to provide shelter from heavy rainfall and a little dense evergreen foliage for roosting; this may be grown in their quarters with no ill-effects since they do not usually damage such foliage.

Food
In the wild, during the breeding season, the brambling will live almost exclusively on insects, taking only some seeds from the ground. At other times this is reversed. The young are reared on insects of all kinds and the adult birds frequently take seeds from opening conifer cones, relying during winter months on beech seeds, where available, and ground seeds during extreme weather. From this, one can see that we have here a species which will readily take insect life, often all year round; they will even catch insects on the wing. The provision of insects in their diet simplifies catering arrangements and ensures a high-protein food intake. Seeds should include screenings or collected wild seeds dried for out-of-season use, plain canary seed, grass, lettuce, all the brassicas, a little linseed, small amounts of maw and gold-of-pleasure and dock. They will rarely eat seeds from growing plants in the wild and, although their

doing so under controlled conditions is a result of failure to provide a totally adequate substitute diet, they do surely benefit from them since they must replace, or partially replace, certain natural foods. Some bramblings will readily eat softfood, especially if its introduction is assisted by its being sprinkled with maw seed or some other favourite seed of the individual bird. Any young birds will require a large number of insects in order to fledge properly.

Breeding
In a mixed collection the brambling must be watched until the individual's behaviour has been noted; the males, and at times females, can become a source of trouble. Not normally of a spiteful nature, they do often become aggressive whilst breeding operations are being carried out. More often than not, in the case of successful rearing, and certainly if they have had full clutches, my birds have had an enclosure of their own. Provide the pair with natural nesting sites by growing an elder in the aviary; when quite bushy in appearance, run a string or wire fully around it at the widest part and tighten it, pulling in the branches to form a tightly knit area; some hens will more readily commence building a nest if a basket containing a moss lining is inserted at this point. When building a nest without the aid of a basket they will require a number of thin birch twigs or similar; I have seen heather twigs taken, also beech and fine dead twigs of honeysuckle. Give roots, moss, dog or cow hair and horse hair, too, if cut short; feathers may also be used but it depends upon individual choice as some birds line their nests entirely with hair; it could well be more natural in their original colder climate to use more feathers for warmth.

The provision of plenty of insect life, mealworms and any other obtainable items of such high-protein value will bring the birds into breeding condition and one will soon hear that raucous call of the male urging his mate to nest.

Bullfinch (*Pyrrhula pyrrhula pileata*)†*****‡

General Description
The British bullfinch is easily distinguished from any other bird in the wild by his jet-black cap and bib, slate-blue-grey back, white rump, black wings and tail, and the rich carmine chest and underparts, which are present only in the male. The female has a brownish-pink chest with browner back, and the young resemble her in colouring. The length is about 16 cm (6¼ in). One can often sex young bullfinches just prior to

their fully entering the moult; then the males can be identified by a slate-grey streak behind each shoulder, which often appears well in advance of the first breast streaks of red.

Habitat
Originally more forest-dwelling than today, the bullfinch is now common throughout almost the whole of the British Isles in woodland, thickets and orchards, the latter much to the professional fruit-growers' dismay due to the terrible depredations on fruit buds. It often frequents gardens in the suburbs and here, too, it will strip fruit buds from many trees; despite this we invariably see many rotting apples on the ground in orchards in the autumn.

Aviary
Again we can use the standard minimum recommended aviary size. These birds are very easy to cater for, but they can be very aggressive and territorially minded. Having once asserted their full authority over the other inmates, however, they often get on with their own domestic duties and ignore other birds. Due, however, to the food requirements of a nest full of young bullfinches – a host of insects in the initial stages of growth, I prefer to give them an enclosure to themselves. The aviary should provide adequate shelter during the winter to shield them from inclement weather; they are hardy birds but live far longer and remain in good health at all times if a dry draught-proof shelter is available to them. Because of their tiresome habit of de-budding so many shrubs and plants, only certain types of growing vegetation can be provided; elder will survive but does not provide what is normally accepted as a good nest site for the bullfinch, so a wicker nest basket should be placed in a dense part of its growth. They will greatly benefit from many kinds of edible vegetables, such as comfrey, cabbage and spinach; hops will also grow, despite the constant pruning, and will provide a source of insect life, and seeds later in the autumn.

Food
Sunflower is a popular seed and a very good flesh-forming food, particularly in the case of nestlings which are over the first 6-day stage. However, in adult birds, such as aging specimens of bullfinch, siskin and redpoll, it can be the cause of obesity. For this reason, it should not be provided too generously to adults unless they are rearing young. Many of our bullfinches breed and successfully rear at 8 and 9 years of age, but too much sunflower, and especially too much hemp, will invariably deter them from doing so. When soaked well, groats will be eaten and

the germ is highly valuable. Once again, endeavour to make use of screenings and many wild field seeds. That wonderful conditioner, dandelion, should be given as soon as the seeding heads become closed after blooming and are on the point of opening to disperse the seeds; at this time there is no better food for advancing condition. Give a mixture of seeds containing soaked pigeon conditioning seed, small pine nuts and safflower, plus smaller amounts each of mung beans and sunflower. Give separately a dry mixture of equal parts British and foreign finch conditioning seed, ground peanuts and pinhead oatmeal, with lesser amounts of teazle, lettuce, chicory and maw. Also provide as many seeding heads from the wild as can be gathered. If available from local farmers, screenings are ideal too. Highly appreciated are the twigs from hawthorn, fruit trees and bushes, just when the buds become swollen and ready to burst open. In fact, as long as the leaves are young and tender, such cuttings as blackcurrant and raspberry will be taken throughout the year. Even my roses, which go unsprayed, provide an odd feed now and again; the birds will take the greenfly and any shoots they can find. I find that a pigeon conditioning mixture, when well soaked and on the point of sprouting, has many seeds that prove acceptable to bullfinches, both for their own use and for feeding nestlings once they have advanced from the insect diet to part-insect and part-grain stage at around 8-10 days. My own bullfinches have, over the years, become fond of soaked household currants; these are given when no other berries or wild fruits can be collected from hedges. I leave them in water overnight, then strain them and include in their seed. During the autumn one can collect many wild fruits, such as hawthorn and rowan, among numerous others, and the bullfinch receiving such attention will moult out far superior to any other; they often give these fruits in small quantities to their young when rearing. Animal and vegetable proteins are the aim in the bullfinch diet and are more important than grain as a basic food. Soaked teazle when they are rearing and during that change-over period from insects to seed will prove a most useful addition to the soaked sunflower. Give sweet apple daily, soaked dried figs, and include perilla seed in the dry mixture or on softfood.

Breeding
Bullfinches require little in the way of nest sites and the fact that they may make only a mere shadow of a real nest is often our fault. We cause so many of the problems which occur when attempting to breed birds; we expect too much adaptation over too short a period. Give the bullfinch adequate thin pliable twigs, plenty of rootlets to choose from and horse, cow and dog hair for lining purposes and they will then build a nest which will hold eggs. When wicker nest baskets are used, the number

of twigs needed will be correspondingly less, but they do still search for such items and, after a heavy shower, when such things are at their most pliable, one can watch these birds carrying them around looking for a suitable site. A wicker nest basket need not be covered from view with evergreen branches, heather or conifer; it merely needs one or two sprays in front of it and some kind of shelter above to stop rain falling directly upon the nest or its occupants.

Northern Bullfinch (*Pyrrhula pyrrhula pyrrhula*)

General Description
This northern bird, or the 'Siberian' as the older bird keeper still calls it, is different only in size and a few minor habits from the British bullfinch. The colouring is certainly brighter than the British bird and it is also larger. I have handled so many over the years that I would say there are two distinct body types: one very long, which although slightly slimmer appears far more so, due to the extra length; and the shorter, very rotund bird, longer than the British bird but also very much thicker set. Some of these birds seem giants when compared to the smaller race and, despite a brighter, more vivid red on the chest of the male, a far greyer back to both sexes, a whiter wing bar and larger areas of these colours, together with the white of the rump, they are identifiable to one and all as bullfinches.

Habitat
Found in northern Europe and Siberia in coniferous forests, their range extends through sparse tree and shrub growth in the very far north. They tend to breed closer to each other than the British bird; even in aviaries they tend to be a little better tempered, or at least more tolerant of their own kind. Even in the north they nest low and require far greater numbers of twigs than the British bullfinch. Some nests I have seen in Finland were immense, almost reminiscent of a dove's nest.

Aviary
The enclosure required to breed these birds should be identical to that recommended for the British bullfinch (page 79). Although the northern bullfinch is used to intensely cold conditions it does not have such wet weather to contend with and therefore a shelter should be provided to protect it, especially from damp. This northern bird still retains its taste for buds and, for this reason, only the plants recommended for the British bullfinch's aviary should be grown in this bird's enclosure.

Food

The food choice of all the races of the bullfinch, from the British Isles across northern Europe, Scandinavia, the former USSR and Asia to Japan, seems to be common under controlled conditions: buttercup seeds, chickweed, charlock, dandelion, dock, dog's mercury, fat-hen, meadowsweet, pansy, sowthistle – the seeds of all these are eagerly sought; the fruit and seeding pips of bramble, rowan, apple and raspberry; buds from many conifers, hawthorn, elm, oak, blackthorn and many soft fruiting shrubs.

During rearing one can see the birds filling their food sacs until their throats bulge; these sacs are never seen or used at other times, apparently shrivelling back to normal when no longer required. All our hens take softfood daily, though I must admit consumption drops during the winter months. Make up a food into a rather stodgy paste and sprinkle this liberally with perilla seed. If this food is new to the birds, persuade them to take it by embedding the perilla into its surface; this will encourage them to taste it. Having studied these birds at close range, and having seen how they literally pump food from these sumps into gaping nestlings, I was amazed by their capacity. The amount of softfood which was taken on each visit was absolutely incredible.

The northern bullfinch does not require anything different from the British bullfinch in the nature of its food supply. When newly imported, mine bred with little encouragement, but the first year they seemed to spend much of their time during rearing searching for foods which were not available in my aviaries; I checked with others who also had these birds in their care and found we were all experiencing much the same behaviour from these birds. In subsequent years they behaved normally or, I should say, appeared to have adapted to our food supplies; apparently some item had seemed to be missing when they bred here the first year; in many instances, this caused high losses among nestlings.

As for all my birds, I tend to soak the seed mixtures. This does not in any way increase their nutritional content but it does tend to make them more easily digested by both young and adult and, more importantly, means one can include other seeds not normally given to finches. Some pigeon conditioning mixtures contain seeds slightly larger than those normally taken, but, in the soaked and sprouting, or near-bursting condition, they are taken by almost all finches and help to vary the diet. In rearing, insects of many kinds, including caterpillars, mealworms and wax-moth larvae, are taken in large quantities; next comes the soaked teazle and sunflower with the mixture of seeds as before, and whatever wild foods can be gathered at this time. Great care should be taken to ensure the cleanliness of wild foods.

Breeding

Nesting sites similar to those used for the British bullfinch should be provided for the larger bullfinch from the north. Their selection of nesting materials does include a greater number of fine twigs, from beech, birch and heather, but anything very thin and pliable will find its way to the rather bulky platform that these birds will construct. A few more will be woven in to form the boundary wall and they will then accept rootlets and animal hair. Coconut fibre may also be used in nest construction.

Such a large proportion of aviary-bred bullfinch hens, particularly of the larger races, are used for hybridising with other finches or canaries that it is essential to breed many more of this beautiful species in the future to offset the hybrid losses; otherwise, with the numbers of hybrid breeders increasing, there will soon be a shortage of the pure strain.

The male can be very possessive of nesting sites and has a reputation for being aggressive towards other finches; for this reason a wary eye should be kept on all aviary inmates of a mixed collection which are in his company while he is in breeding condition. Of course there are numerous examples of cases where no trouble occurs. I have bred from four and five pairs in a 5 m (16½ ft) by 13 m (39¾ ft) aviary without a loss of feather; all were genuine northern bullfinches and each had taken up a territory since early December of the previous year. I believe the secret of success on those occasions was the neutral feeding areas, with no nesting sites adjoining them, and the introduction of all the pairs at the same time.

I have over the years acquired a great faith in early pairing of birds and their introduction into selected breeding enclosures. It enables birds to build up a confidence in their keeper and familiarise themselves with usual sights in the neighbourhood: dogs, other birds, general movement of the passers-by. The sound of the male's rather sad and melancholy single call note changes with the season's longer days. The song proper, when it commences, sounds almost as though the male bullfinch is chuckling with delight; the low attractive bubbling, although often made by a hen, usually announces his attentions to his mate, as he swings his tail from side to side, ducking and bowing whilst sidling towards her.

The eggs of this bullfinch are slightly larger, with markings much the same but with a tendency towards heavier reddish-brown spots. Incubation takes about the same time and yet seems to occur a shade sooner and I believe the growth of both chick and feather is advanced in this larger variety. To confirm these theories, whilst all the time one is endeavouring to introduce all the body-building foods one can, is a little difficult, and far more controlled research, with numerous pairs, is essential to prove or disprove them.

One of the rearing foods first accepted by these birds was sowthistle which I have seen in Finland, Scandinavia and on the border of the former USSR, so it is quite probable that they are already familiar with it as a food for this particular purpose.

Chaffinch (*Fringilla coelebs*)[†*****‡]

General Description
Reputed to be Britain's most common bird at one time, far fewer, sadly, may be seen these days. The sight of one in his breeding plumage is certainly an admirable picture and a pleasant experience. A brilliant slate-blue crown, chestnut back fading to a beautiful shade of rich green on the rump, the dark reddish-pink chest and very dark wings with their white patches shown off to perfection allow this bird to compare well with many tropical species. Both sexes are about 15 cm (6 in) long but the hen is far browner in all areas. Despite this she is a very attractive subject when viewed in really good plumage and overall condition. The winter plumage of the male is less exciting but nonetheless attractive with reddish-buff head and nape; the back is merely a pale chestnut hue.

Habitat
The chaffinch of northern Europe extends into Siberia (see northern variety, page 86). It is a bird of deciduous woods and the northern spruce woods, but present in gardens and parks, even in built-up areas. Well-grown woodland still has the more dense populations.

Aviary
The standard minimum enclosure will suffice but, where space is not at a premium, give a longer flight. The shelter need be little more than 0.5 m (1 5/8 ft) of roofing with a little cover at the rear and sides for added comfort. The longer flight will be found an asset when the breeding of these birds is attempted. The male is quite a boisterous wooer, such that I think at times I would use a more colourful description if I were a hen chaffinch! He can, when in top breeding condition, continually chase the hen, singing all the time, until she agrees that sitting on eggs is far less arduous a pastime.

Food

Feed as recommended for the brambling (page 77–8), its nearest relative. Many seeds are taken on open ground, fallen beech mast being a favourite in all regions where it exists. The more common weeds, persicaria, chickweed and their like, form the bulk of its diet. At one time after the threshing in farmland, when the waste was thrown in heaps, the number of chaffinches eating in the area was almost unbelievable, hence their name. During the breeding season, the diet of the adults, although containing many seeds, is basically insectivorous; insects will be taken from the ground, trees, and even on the wing, and moths, caterpillars, grubs and small beetles of all kinds are taken for the young. In aviaries we must, if hoping to rear full nests of young, see that they receive similar diets. The seed contents of the mixture, which must always be given daily as a basic requirement, should contain many of the brassicas, grasses, lettuce, plain canary seed, some linseed and a few grains of hemp. Fallen seeds from wild-collected weeds will be taken at all times but this is one of the finches that rarely take seeds from a growing weed and seldom from such plants as the groundsel and dandelion. In an aviary they will do so but only very spasmodically.

Breeding

Having given the pair of birds an aviary to themselves, or, if in a collection, kept an alert watch for bullying of smaller finches by the male, we can do little apart from providing a high-protein insect intake, watching nature take its course and see these birds come into condition. The male will chase the hen quite violently at times, yet, when accepted by her, he will be one of the most devoted of mates, offering her mealworm after mealworm whilst she broods her clutch. The nesting site usually ends up being self-selected, as they tend to show complete disdain for nest baskets. It matters not if it is a hand-reared bird which has never seen another chaffinch apart from its mate; the nest is a work of art, just as one would expect a wild bird's nest to be. This sort of inherited knowledge can be absorbing to study in many ways, in all birds. Materials for nesting should include mosses, spiders' webs, as wide a variety of animal hair as possible, a few feathers, small pieces of tissue paper, fine grasses, thin rootlets and lichen; the nest building is an art in itself and well worth using the binoculars to observe. The mating behaviour, the varied courtship postures and the response movements and calls make a fascinating study when time permits. Occasionally a hen will take quite a long while to make up her mind about nesting and the making-up of a few more sites will often give her the required incentive, but with this species so much is the result of the courtship;

the male's actions quite visibly act as a key to the female's acceptance and nesting.

With the arrival of young, the need for insects must be fulfilled or they will shrink and be lost: the first 5 days are the most vital, as with so many young birds dependent upon animal protein for their existence, so provide all possible from the small insect world.

Chaffinch (*Fringilla coelebs*) Northern Variety****†

General Description
The northern variety of the chaffinch differs only in appearance. It is larger, I would estimate by as much as 1 cm (½ in) in length. Compare this increase, in a bird measuring 14-16 cm (5½-6½ in) long, with a similar percentage increase in the height of a human being, and the size variation becomes appreciable.

It has correspondingly increased areas of colour; certainly the white wing patches can be very noticeable and so can the deep plum-red of the breast, the area of green on the rump and the rich back colour. Set one of each, side by side, and one can very quickly spot the greater colour regions of the body. With the vast movement of chaffinches annually, breeding ranges of immense proportions and their general movements, there is obviously no distinct dividing line but when comparing a bird taken in the former USSR with one resident in the British Isles the differences are immediately recognisable, and even many northern European specimens exhibit disparity in size.

Naturally, the habitat varies and, since the bird lives on the edge of the wooded steppes just below the tundra areas, the vegetation, and the availability of nest materials and food, must, to some extent, change with the terrain.

Blue Chaffinch (*Fringilla teydea*)

General Description
This is another chaffinch showing distinct variations and there are specimens from North Africa and Spain which differ again, but I have never handled them. This all-blue chaffinch or, to be more exact, slate-blue male and dull greyish hen, is quite unmistakable; it completely lacks the pink, brown and chestnut markings of those seen in northern Europe. Housing, feeding and breeding will be identical, each becoming an almost insectivorous creature with the arrival of the breeding season.

Nest materials are very similar, varying slightly from one adopted habitat to another. Nest sites, too, are only marginally different: where chaffinches live in olive groves instead of in hawthorn or other trees, they will use them from habit.

Common Crossbill (*Loxia curvirostra*) and Scottish Crossbill (*L. c. scotica*)****

General Description
These birds are here classed together since it is only through continued and long association with the pine that the Scottish crossbill has developed its larger bill and this is the only recognisable difference between them. When describing the adult male as red, the immature male as orange-red and females as olive-brown, one can easily be accused of over-simplification. A really young male resembles a young greenfinch, slightly stouter but very speckled; at the first moult he does acquire some orange but, at this first stage, he can be confused with a well-coloured adult female which has a bright face, chest and rump. This, however, is as bright as the female ever normally moults. The adult male is a dark red on head, chest and rump, with the wings and tail dark brown. The young males often show a mixture of shades from yellow through orange to orange-red and this, to a certain extent, depends upon their age and the availability of fir cones during their nestling stage and first moult. The adult female is green but showing a greyish clouding, has a yellowish-green face and rump, is greyish on the belly and very streaked if a first-year bird; she never acquires any true red, merely a shade of yellow-orange at the best; the denser red shades are found only in the males of all crossbill races. Both are 16½ cm (6½ in) in length.

Habitat
The common crossbill is a lover of the coniferous woods and forests and the range of this bird extends over much of Europe, Asia, North America and Canada. The cones of spruce and pine provide the basic food requirements, although buds of conifers are also taken when cone crops fail. Towards the end of the eighteenth century, when coniferous trees were planted extensively, and then with present-day re-afforestation providing a new habitat, it is not surprising that, some years after irruptions from the Continent and Scandinavia, vast populations of crossbills built up in the British Isles. They are now quite a common breeding bird; fir cones pulled to pieces beneath pine or spruce usually

indicate their whereabouts. The cones are not damaged in the same manner as the squirrels' chewing of them; the crossbill neatly extracts the seed with his beak, which is crossed at the end in a scissor fashion and well adapted for its purpose.

Aviary
Obesity is an enemy of the crossbill in confinement and one must, if wishing to breed from these birds, provide ample quarters for flight exercise; a minimum would be 2 m (6½ ft) by 1 m (3¼ ft) and 2 m (6½ ft) high. If a longer enclosure can be provided so much the better. When in the wild, this bird loses much energy in food collection and providing a dish of fattening food, available merely for the taking, does not assist in keeping it in the best condition for breeding. To start with, it is better to place food at one end of the enclosure and the drinking and bathing water as far as possible in the other direction; at least this will ensure the maximum amount of flight being taken. The crossbill is a large strong finch but can put on weight very quickly with a subsequent loss of general health and virility. Apart from roosting perches of conifer branches fairly high in the enclosure, their perching need consist only of one horizontal fir branch at either end of the long aviary. There is no need of shelter, except for a slight covered portion of the roof. The crossbill family does not care for too much rain or damp, despite its hardiness against cold weather.

Always ensure cleanliness of perching arrangements; study the roosting area for soiled perches and remove them at once, replacing them with new clean branches, fresh from the wild if possible. My reason for this last piece of advice is that the crossbill invariably roosts in the same site each night and fouls nearby perches. A microbe which breeds in these accumulated droppings will eventually find its way beneath the leg scales and cause scaly-leg; I have always treated this with a mixture of 5 per cent salicylic acid and 95 per cent Vaseline, rubbing it into the infected leg once a week for 2 weeks; this is normally all that is required to rid the bird of the white crusty build-up but, if clean new perches are not provided, it will soon return. Always keep this mixture away from the bird's eyes and wash your hands after applying it.

The aviary will take on a far more pleasing appearance to the observer, and certainly be appreciated by the inmates, if it is planted out with conifers. Stunted pines are best if the tops are removed at about 1.5 m (4⅞ ft) and then kept trimmed during the winter months to prevent them damaging the wire-netting roof. A certain amount of chewing is unavoidable, but if growth at each end is fairly dense, and cones are collected and offered regularly, only a few branches will die off.

Food

Now that various sizes of pine nuts are freely available commercially, bird keepers need no longer be dependent upon somewhat spasmodic supplies from health food stores. This valuable food can be included regularly in larger quantities for most of the finches. The majority prefer the smaller variety; these I soak with, and include in, the basic soaked mixture. At one time I used to bring pine nuts from abroad, and also pirella seed, on every possible occasion because, in those days, both were scarce here in the British Isles.

One should still collect fir cones for crossbills; apart from being a natural source of food, they keep these birds active and working, which lessens immensely the time they otherwise often spend between meals just perching. Even a few cones a week will occupy a bird for many hours. Give lavishly the usual assortment of seeding vegetation, sowthistle, chickweed etc. Do not offer too much in the way of seeding pods of cabbage or other brassicas because the droppings will become very loose. Even seeding grasses, dandelion, ripe rhubarb and dock seed when both are reddish-brown, seeding hops, berries of hawthorn and rowan are consumed by many finches, but I have a suspicion that the food selection varies with the origin of the birds in question. We always tried to provide our birds with a mixture made up as follows: two parts small pine nuts and one part each sunflower and safflower, together with the pigeon conditioning mix, plus a little each of mung beans and buckwheat. Recommending screenings today, when apparently only a minority of people can manage to get them, may seem odd, but such seeds can be valuable as natural feeding stuff. Failing this wild food, offer a dry mixture made up of British and foreign finch conditioning seed, ground peanuts and pinhead oatmeal. All our birds have been extremely fond of beech mast – those sweet-tasting, three-sided little nuts which can almost be shovelled up in season; they are of very high food value, rich in fats and proteins and the botanical name *Fagus* is derived from the Greek word meaning 'to eat'.

Many of the crossbill races can be persuaded to accept some kind of softfood; apart from proving an asset when rearing, it is also a useful medium in which to introduce colour foods, which the males should receive at each moult. If a softfood is given as a thick, rather stodgy paste, well sprinkled with pirella seed, it is not normally too long before the birds also eat some of the food itself. To begin with, just thumb the seed into the surface of the food. Do remember to keep the crossbill sexes apart when colour feeding or some hens will come to resemble males so closely that errors of identification can occur among birds kept in large enclosures.

Breeding

Shelves made of conifers or groups of conifer branches bunched together to form individual sites are usually acceptable to these birds when in true breeding condition. The nest requires some twigs of birch, heather, beech or similar fine twiggy pieces up to 15 cm (6 in) in length; these will be built into a platform on which the nest is constructed. Offer moss, dry grasses, strips of bark fibre; animal hair from cow, horse or dog brushings will prove ideal. A few feathers are sometimes taken for lining but I have found nearly all aviary nests lined only with hair. Nest baskets, whilst not exactly ignored, are seldom if ever used for their intended purpose; most of those I ever offered were torn to pieces and strewn around the aviary floor; I believe two only were ever used by birds for nesting operations.

Although not able to guarantee the behaviour of these birds in a mixed collection at all times, I have had pairs nest under such conditions without any damage to other inmates. After the occasional open-beak threat the pecking order seemed firmly established; other birds avoided the nest site. Mine have all appeared quite docile specimens. Even so, for successful breeding, an aviary to each pair should be the general aim.

The eggs are bluish-green with brown and brownish-purple spots, and the most I have found in their nests has been four. The young remain in the nest longer than most finches and I have also managed to rear them under foster parents. On that occasion the foster parents were greenfinches which had been hand-reared by my daughter when she herself was very young; as a child, hand-rearing fascinated her and, each year, I let her rear a nest of greenfinches, usually those removed (normals) when I was concentrating upon a mutation of greenfinch plumage. These birds invariably received a daily quota of mealworms and grew to like them so much that the daily ration continued into adulthood. A pair of these birds hatched and reared three young crossbills without the aid of any pine seeds or other normal crossbill rearing foods; daily supplies of mealworms and the usual wild foods and soaked seeds one associates with greenfinch rearing were used.

Over the years I suppose I have kept about 40–60 various crossbills, releasing the majority after studying them. They are by far one of the most interesting finches. I have examined hundreds in the field, mostly when I was younger and wished to see if the crossing of the beak tips bore any significance to sex or the type of cone fed upon etc. The beak tips can cross either way, regardless of these factors, and there seems no uniformity in this respect.

Parrot Crossbill (*Loxia pytyopsittacus*)[†]

General Description
Larger than either the common or two-barred crossbills, at 17 cm (6¾ in), this bird has a beak of a far heavier type, hooked more like that of a parrot and wider in appearance from above, front and below. When observing the skull in a skeleton, the difference is quite remarkable; the rest of the skeletal frame is slightly larger and disproves the theory that greater body weight is due merely to flesh. The colour, seen from a distance, appears more or less the same as that of the common crossbill but, when in an aviary, one can easily discern a more greyish head in the female and the whitish tips to the male's greater coverts. The call is deeper, more of a 'chop chop' in comparison to the 'chip chip' of the common crossbill. Probably due only to this beak of greater dimensions, this bird does look very parrot-like about the head, and all crossbills tend to peer at one in a rather parrot-like fashion, sideways or as though looking over their glasses. This bird's behaviour is even more pronounced in this respect and it often seems to be pulling faces at its observer.

Habitat
The parrot crossbill is a bird of Scandinavia, the former USSR and Finland, breeding in these countries and to a far lesser degree in Poland and eastern Germany and apparently a very rare visitor to Britain. Since they are so similar to the common crossbill when observed in the field, there could easily be more unrecorded visitors. The crop of pine cones is the basic diet, with a supplement of buds, insects and, during the late autumn and winter, the berries of *Sorbus* and other fruiting trees.

Aviary
It is always difficult to make hard-and-fast rules with regard to any livestock. Breeding is frequently attempted under the most unsatisfactory conditions, sometimes even proving successful, but I would never consider giving a pair of these birds an aviary smaller than 3 m (9¾ ft) long by 2.5 m (8⅛ ft) high, with a width of anything less than between 2 and 3 m (6½ and 9¼ ft). They tend towards laziness if food is always available and they easily develop a rather bad habit of perching for long rest periods between feeds. Provide perches at each end of an enclosure, if possible with food one end and water at the other. By all means plant their aviary as near naturally as one can; stunted pines are ideal for this purpose. Endeavour to give perches from conifers, and, as often as possible, offer fresh pine branches with many large tightly closed full

cones. The work entailed getting the natural seeds will not only ensure an active life but also provide the natural food essential to this bird. Due to high resinous content of their wild diet they are avid drinkers and, even when snow and ice abounds, they will still take their daily bath, so do ensure an available source of clean water.

Food
Since the natural diet is almost an all-year round one of fir cone seeds, especially from the pine, the provision of this food is always greatly appreciated. Sunflower is not harmful to them and so can be given in reasonably large quantities. The tendency of all crossbills to eat their fill then sluggishly rest until again hungry must always be remembered, and it is not advisable to over-do the supply of this seed. Place a seed mixture consisting of sunflower, groats, a little niger and a good-quality pigeon conditioning seed in water to soak for 24 hours; strain before feeding and offer in a deepish bowl; the crossbill often forms a liking for a particular seed and will scatter everything else to the four winds. Some food will remain if a deep seed container is used and the work involved will prove beneficial.

Almost all autumn berries will be taken; the buds of fruit tree prunings or hawthorn hedge cuttings are always welcome as, too, are many seeding grasses, sowthistle, dandelion, chickweed, sweet apple and the pips from apple cores; from fresh supplies of pine branches they will take every bud with their natural yearning for the taste of pine. Cut up mealworms on their seed supply and, when they are observed taking them readily, merely pinch the heads of the mealworms; as soon as these are eaten, offer the live mealworms. Once accustomed to this conditioning food, they will remain fit and healthy for many years. When rearing, and crossbills will certainly breed when really fit, they will need many mealworms for their nestlings. It is mainly these, buds and the soaked sunflower that will then form the basic diet of the young.

Some crossbills will take to softfoods and, if these are mixed with milk as the liquid base, it is another way of introducing animal protein in plenty to their diet. I have seen birds which eat avidly of bread and milk, others that take Farex (Glaxo) baby food and some that take to an insectivorous mixture if it is covered with cut-up mealworms before it is proffered.

Breeding
As would be expected, these birds nest on pine branches and, knowing this, one should put one's mind to fabricating either shelves of pine branches or clusters of them firmly secured in a position near the aviary

roof. Remember to provide ample conifer above the site to shield it from excess sun or rain; as a bird of the north, it finds excess heat unpleasant, so give plenty of shelter.

Breeding condition is reached very early in the year; in fact, this is a bird which nests almost anywhere between December and June.

The parrot crossbill's nest is a much heavier structure than that of the other crossbills. It is mainly made up of twigs which form a platform and lower wall, not merely for strength but to form a barrier against cold. Other materials include grass, moss, bark fibre, lichen, animal hair and feathers. The whole is a clumsy-looking nest but it is highly suitable for the wild habitat. The eggs are bluish-green, heavily spotted with brown and purple. Resulting young will remain in the nest for about 21 days.

Two-barred or White-winged Crossbill
(*Loxia leucoptera*)**

General Description
A very striking effect is achieved by this bird's plumage which is very different from that of other crossbills. It is a little smaller than the common race, at 15 cm (6 in) and a darker and brighter red, but of course it is the double white bar on each wing that makes it so attractive. The bill is smaller and neater looking, but is nevertheless most useful, for the normal food is the seed of the small larch cone. The bird is much daintier in appearance and less clumsy when feeding, even in the natural state. I have always been amazed at this bird's hardiness as its range takes it further north than the common crossbill. I have only managed to keep four of these birds and, of these, only one hen nested and laid eggs. Professor H. B. Tordoff (USA) bred hybrids between this and the common crossbill which proved fertile.

Habitat
In the Old World this bird belongs to northern Asia and the north-east of the former USSR and breeds there among the Siberian larch forests. Although it feeds extensively on larch seeds from the smaller more delicate cone of that tree, this crossbill also takes many insects and berries which contain pips or seeds, the rowan being a favourite. As would be expected, nests are made in the larch and are not quite such bulky constructions as that of other crossbills although basically the same materials are used: fine twigs, grasses, moss, lichen and hair.

Aviary

If fortunate enough ever to acquire a pair of these birds, give them a long aviary of at least 4 m (13 ft). I have noticed that this race of crossbill flies far more, and in much lighter fashion, than the others. A height of 2 m (6½ ft) and the same width would be ideal as a minimum. In all other aspects their enclosure could be a typical crossbill aviary, planted with pine or larch with food at one end and water at the other. A small roofed portion should be provided to shield them from heavy rain and too-damp roosting facilities. This bird never chewed the aviary timber in any way, whereas occasionally one of the other races made the steady destruction of the timberwork a lifelong project.

Food

I gave my two-barred crossbills a little more niger than the other crossbills but, in all other respects, the seed diet was identical, invariably soaked during the months when there were no frosts and supplied, sometimes sprouting, in this softened state. All other foods, such as dandelion and sowthistle, were taken immediately they were offered. I think they take more fruit quite naturally, not just when the cone harvest fails and I know that mine were particularly fond of apple pips; they still ate much of the apple flesh but, when a core was offered, they immediately set upon it to reach the pips within.

Breeding

If the nesting pattern of my birds each year was anything to generalise upon, I feel that many two-barred crossbills must be lost in the wild to predators. The nest was made on the outer branches on each occasion, in full view of anyone walking along the adjoining path; these birds have little or no fear of human beings. To aid their nesting operations I erected quite elaborate conifer shelves, built in about three tiers, which they almost completely ignored. As if to spite my efforts they built on the very outer edge of one branch. They require fine twigs, mosses, grasses and hair. My birds never seemed interested in feathers for nesting.

At the time of keeping these birds I also had a great surplus of wax-moth larvae and I found that these were often taken in preference to mealworms, despite the latter having been gradually introduced by being cut up upon the daily seed mixture.

On the first breeding the male of my pair was not really fit as I had only just acquired him from a continental dealer and the eggs proved clear. In the following years the hen nested and laid but without a male being present and, much later, when I learned of hybrids having been obtained from pairing these with the common crossbill, I wished I had also attempted this. Even if not for the sake of achieving hybrids it would have been very interesting to see how many generations it took to breed back to almost pure two-barred crossbill.

Citril Finch (*Serinus citrinella*)****

General Description
This rather greyish-green bird is rarely seen as an aviary subject, probably because it resides in the high altitude regions of central Europe. The first impression is that the green is very yellowish but this, upon closer inspection, is found to be misleading as there is much slate-grey in the plumage of both male and female. The underparts are not striated, as in the siskin or serin, and this exhibits the yellow to advantage. The adult male and female differ mainly in the amount of grey; in the male this is predominantly around the nape, neck sides and even the top of the head whereas the female has a greyish, slightly more streaked body, with much more grey in the chest area as well as on the nape and head. The wings and tail are black, or almost so, and there are two dullish-green wing bars visible on close inspection. The length is just below 13 cm (5 in) and the bird is of slim build. The juvenile is brownish with shades of green only on the under-body; it does not have the wing bars and is often mistaken for other birds.

Habitat
Fairly high mountainous areas among the conifers would be a typical sighting place for this bird where it feeds directly from the cones. Some of these birds remain in the Swiss alpine regions all the year but many migrate south to France. Grasses seem the most taken seed, but dandelion and many others provide variety and I am not convinced that a number of insects are not used in the rearing of their young, although they do offer nestlings many seeds. I have never undertaken crop inspections in the wild. I have found a number of nests high in the trees, resembling those of the goldfinch or siskin in that area.

Aviary
Due to this bird's almost fixed choice of high nesting site, one should consider making a special effort to afford extra height in the aviary. Where breeding is the main aim, offer a high flight up to 3 m (9¾ ft). This is frequently not possible but a gabled roof would provide this amenity: a 2 m (6½ ft) length by 1 m (3¼ ft) width should prove ideal. Bearing in mind the bird's natural habitat and choice of conifer,

endeavour to provide spruce cones and fresh branches. Only when this bird feels really at home in its surroundings will it nest. The ground area of natural earth should be well planted with seeding grasses, comfrey and dandelion.

Food
The seed mixture for this bird would ideally be screenings containing plenty of grasses. Not too much niger should be given or it tends to become as addicted to this as many other finches; the large seeds can be forgotten unless the soaking of seed is undertaken, although I have seen citril finches on occasion even work on the large sunflower seed and remove the contents quite easily. Bunches of seeding grasses hung in various sites can prove entertaining as well as beneficial to the birds. They will hang from the stalks in many positions, taking from the ripe harvest of seeds, and only when the supply is exhausted revert to man-mixed varieties. Where screenings are available mix in some plain canary, a little niger, small quantities of rape, linseed, gold-of-pleasure, maw and pinhead oatmeal. Millet will also be taken and I have had birds which took to this seed very quickly, the consumption amounting to almost a third of their daily intake of seed. Some specimens take to a softfood readily once maw has been sprinkled on top and acquire a taste for this food. In aviaries, they do certainly take many insects and I have carried out crop inspections among young and adult birds to find that all took greenfly whenever it was available.

Breeding
The provision of plenty of conifer branches and rich feeding, together with an aviary well shielded from the direct strong sun rays seem to receive this species' approval for breeding. I have seen hens gasping and raising themselves off the nests in an effort to keep cool; they are a high altitude species and I have always kept them in conditions as near to their natural habitat as possible. Of those birds I have kept, very few have nested in mixed collections; perhaps they suffer from a feeling of insecurity when among larger species, for they are rather nervous by nature. On at least four occasions in mixed collections, I have had eggs from them hatch, but it would seem that the competition for the rearing food was too great for no young fledged. Those pairs which reared successfully were, on each occasion, housed in their own breeding pens and seemingly did so without any undue effort. They were supplied with the same foods as previously: large bunches of seeding grasses hung from the roof, chickweed, sowthistle, greenfly, wax-moth larvae and even ants' 'eggs'. This bird is well worth breeding, although I know of

occurrences, in the hands of very accomplished aviculturists, when they have appeared to be short-lived. There is certainly much to be learned by studying this species; I feel that when they nest so readily, incubate with no problems and then rear their young when housed on their own, they should be bred under controlled aviary conditions as frequently as the common serin or siskin.

Snow Finch (*Montifringilla nivalis*)

General Description
This alpine species from mountainous central European areas is larger than the snow bunting. It is indeed very surprising when one holds them in the hand to find just how large and bulky the bone structure is; yet the bird is rarely plump, just solid. They are very striking specimens and very tame, with little fear of human beings. The male, in summer, shows a rich chocolate-brown back, grey head, black throat bib and whitish belly and underparts. The wings appear very white in flight. On the ground it will be seen that the primaries are black and there are black central feathers to the tail with white on each side. This vast amount of white is very conspicuous in flight, giving the bird the effect of a rather large butterfly. The female is much more dull, having a browner head and less white in wings and tail. The bills of both sexes are yellow, turning black in the winter as the spring breeding condition builds up. Juveniles are rather dull in appearance, but on the whole this bird is a remarkably attractive species. Feeding in the winter months is often in the close vicinity of human beings, their presence indicating a source of food for the birds. They frequently feed in quite large flocks, more often than not in the snow where they make an attractive sight. The length is 18 cm (7 in).

Habitat
In its alpine habitat this bird's tameness is its most striking characteristic; it seems to avoid trees and remain on open ground wherever possible. The favoured foods are seeds and insect life as they become available but the winter months, with their scarcity of food, bring this bird nearer and nearer to man and the scraps of food and crumbs he provides; many tourists in alpine resorts find this bird's tameness a fascinating item of conversation as it feeds almost at their feet. It seems happiest in the summer high among the rock crags, but it still remains not too distant from human habitation, frequently roosting in a hole in building walls or under the wide jutting eaves.

Aviary

Despite this bird's tameness, one should provide ample aviary space, 3 m (9¾ ft) long by 2 m (6½ ft) wide and high, so that the enclosure gives ample room for both exercise and breeding operations, together with an adequate flying area so that one may appreciate its plumage to the full. It is difficult to build an enclosure which simulates the bird's wild habitat unless it has a back wall; if this is available it is then a simple matter to build a second rock wall against the first, leaving nesting holes in suitable places for them to choose from during the breeding season. Quite apart from serving as ideal and natural nesting sites, this rock wall will give the birds a perfect setting, and a few juniper bushes and heathers planted in the enclosure will provide an almost natural habitat.

Food

Whilst insects are almost the sole food during the warmer months, there is no doubt that seed and softfoods will be readily eaten during the winter. I have always offered a choice of both to this bird and, although I have kept only a limited number, I have had ample opportunity to study the wildlings. Once again I would suggest screenings as the base of a seed mixture, to which I would add plain canary seed, a little rape, small millets and pinhead oatmeal. I would give seeding grasses as well as the normal wild foods, such as chickweed. A few maggots can be given daily and the odd mealworm from the hand or placed nearby will encourage the attractive friendliness of this species. During the summer months one can easily increase the intake of livefood and many of those insects which enter the enclosure will be taken.

Breeding

Where breeding is attempted, or looks likely, offer plenty of fine dry grasses as the lining of their nest is composed of animal hair and feathers. It is rather a clumsy nest but when one sees the actual size of the adult bird, it can easily be visualised just how much space four or five youngsters will need when fully grown and about to fledge. I have known this bird to use a nest box – the half-open-fronted type – but for those wishing to achieve as near natural a success as they are able, rock crevices and holes are the sites to strive for.

During rearing these birds will require much livefood. I have seen most forms taken; the simplest to culture in large numbers are locusts, mealworms and wax-moths. On the few occasions when my birds received wood-ant 'eggs' they seemed particularly attracted to them and, knowing how these creatures often abound in mountainous regions where fir trees are plentiful, I have little doubt that they take them when available in the wild. The eggs are whitish and, from wildling observations, there are frequently two broods. Both sexes appear to take turns at incubation and rearing the young is also shared.

Goldfinch (*Carduelis carduelis*)[1*****‡]

General Description
The European goldfinch must be one of the most well-known birds for colour: its many-hued progeny when paired with the canary or other finches, all retain a resemblance to the very popular red-faced bird. It is true that there exists wide variation in size and even in brightness of colour throughout Europe but until one wanders to the far side of the Urals no great plumage pattern changes occur. It measures almost 12 cm (4¾ in) in length in the British Isles and, whilst it decreases in size in more southerly countries, the Siberian or Archangel variety, as those enormous specimens with a hoary-like frosting on their black-topped heads used to be called, show a marked increase in stature. The goldfinch has chestnut back and flanks with slightly paler breast markings (called tannings by the aviculturist); the under-belly and cheeks are white and the black crown tapers till the points almost meet the throat. There is a brilliant red facial 'blaze' or mask; the tail and wings are jet-black, the wings having broad daffodil-yellow bars and the feathers of the wings and the underside of the outer tail feathers are tipped with white. Some females are as brilliant as a mediocre male but usually they are a shade duller.

To see a flock of these birds descend upon a patch of thistle or dandelion seeding heads is a wonderful experience for the bird lover. With notes tinkling like bells and the wings flitting to exhibit the deep yellow bars to best advantage, in contrast to the red of the mask, it is indeed a bird worthy of all the colloquial and local flattering names bestowed upon it.

The long pointed bill is an ideal tool for obtaining the seeds of the thistle. Both male and female will cling to the ripened head and work at the tightly packed seeds though the male, with a slightly longer and stronger base to his bill, is a little more proficient at this task than his mate.

Habitat
This bird can be found on woodland edge, field, agricultural land and wild moorland, and often in many tree-lined avenues and gardens in suburbia. The orchard, often frequented for nesting purposes, has in

many areas been forsaken and preference shown for the odd tree in a garden, or even a chestnut growing along a busy road. The dandelion and thistle still remain its choice of foods and it inhabits almost any region where these abound. Since the goldfinch breeds over very nearly the whole of Europe, the habitat does vary with the country in which it lives and breeds, or to which it migrates to during the winter months.

Aviary
There are various methods of housing goldfinches. People who want to form a true strain will methodically pair their birds for excellence of colour and shape; others who may wish to study them in a group may colony breed them, a number of pairs to a large aviary. This is certainly a bird which can be housed more than one pair to an enclosure, provided, of course, that the dimensions are generous enough to permit it. For one pair the enclosure would be the standard minimum, the length not below 2 m (6½ ft), the height the same, and the width not less than 1 m (3¼ ft): for two pairs add a further 50 cm (20 in) to the width; if considering more than six pairs as a breeding colony then one should aim for quite extensive quarters. Endeavour to keep a large neutral feeding area when housing a number of birds in one aviary; there will be far less bickering and the long-term results will be appreciated.

Plant a goldfinch aviary with a number of roots of comfrey and dandelion. These birds can be destructive with regard to shrubs and flowering trees, so hard-wearing items such as elder, hops, or willow can provide perching and cover, together with a generous supply of livefood in the way of aphids during the months when they are most welcome. The goldfinch does not enjoy damp quarters or draughty areas so provide warm dry roosts and plenty of roof shelter, but allow the maximum of roof light to enter their aviary. They require no winter heating but try to protect them from rainfall and draughts; hard frosts when their crops are full do no harm at all. A strip of plastic sheeting around the outer edges of the roof, perhaps extending for an equal distance down the sides of the aviary, will ensure dry roosts, safety from cats and owls and shelter from direct winds.

Food
When referring to pigeon conditioning mixture, I use Haiths which is a rich and varied brand. It should be soaked for 24 hours, with additional seeds such as small pine nuts and a few mung beans. These birds also receive a dry mixture composed of both British and foreign finch conditioning seed, ground peanuts, teazle, chicory and lettuce.

This is one of the few birds that does not appear to get too fat on a rich diet, though the food supply must not be exceptionally rich or the liver and kidneys can suffer. I still like to keep to screenings when possible, with the inclusion of a wide variety of seeds, as used in many standard mixtures. One can even add a little hemp. The soaking of seed allows a much wider variety to be offered. Some seeds, such as those in pigeon conditioning mixtures, are often too large and hard for many finches without prior soaking, straining and sprouting. For these birds I would add equal parts of the following to a good clean screenings mixture: teazle, millet, rape and niger, then a half part of each hemp, maw, linseed and pigeon conditioning seed; lettuce seed and dandelion is also suitable. Many goldfinches will take softfood once introduced to it by a generous sprinkling of maw seed on top of it and many are quite content with a baby-food base to which one can add many further attractions, such as gold-of-pleasure seed and maw. The birds which do take freely of such foods are usually the best rearing ones when breeding is in progress. Occasionally, when a goldfinch begins to lose weight and nothing seems to combat the problem, an introduction of mealworms into the diet, cut up on the seed, will remedy the weight loss and bring the bird back to good health in a very short while. I have even experimented with the use of milk at this time, in lieu of water, and found this also a great help, assisting in the replacement of body flesh, particularly over the breast bone.

Breeding

One of the easier finches to breed in single pairs, trios or on a colony system, these birds will breed successfully if housed and fed well, with adequate additional food for rearing and extra space when a number of pairs are involved; the food supply should be situated in a neutral feeding area. Elder, willow and hops will provide good cover for nesting sites. Baskets with a few small branches of conifer secured around them or even clusters of birch twigs surrounded by conifer can be fastened within the growing elders or willows, preferably at near-maximum height in the breeding aviary. Natural nests will certainly be constructed where good sites exist but none of the above-mentioned growing vegetation provides ideal sites for this purpose. With these birds' destructive habits of eating buds of some growing trees and shrubs, it is seldom that a plant develops sufficiently thick growth for natural nesting and clusters of evergreen should be added. Nesting materials are fine dry grass, moss, rootlets, plant down, animal hair, small feathers and even little tufts of cotton-wool; the last mentioned should be given only in a well teased-out condition, in small fragments. The eggs normally number between four and seven; they are pale blue, finely streaked and spotted with dark purple or red.

With the arrival of nestlings, aphids, chrysalids and mealworms should be provided in plenty. The adults are quite willing to feed these items to their young if well used to them in advance and not suddenly given them for the first time when young hatch; such foods will give the young a very good start in life. Give sowthistle and chickweed but when comfrey and any brassica, even the rape plants growing from wasted seed, are growing in the enclosure, the crops of the young will frequently appear almost black, with the contents showing through the skin.

The normal soaked-seed mixture should now have generous additional proportions of soaked teazle, hemp, safflower and small sunflower seeds; where colony breeding is being practised there will be extra demand and consumption by birds other than the breeding pairs. If the goldfinches are housed with other varieties of finches it is advisable to forget the hemp entirely for fear of obesity, with subsequent egg binding, or possibly respiratory ailments. The only exception is where linnets are the only other finches housed with the goldfinches. Keep on feeding the soaked-seed mixture, ensuring that continuous supplies of both greenfood and insect life are available right through the moult of the young birds.

Over the years I have many times experimented with rearing goldfinches and I can state emphatically that my best-reared goldfinches, native or of the large race, were reared entirely on soaked sunflower, teazle, mealworms and comfrey, with hardly anything else except a little human baby food mixed with milk and placed on the food dish.

The collection of wild food when one has birds rearing is a good thing provided that the source of supply does not suddenly dry up when least expected. Should this happen, the sudden need for change at this crucial stage is sometimes fatal. It is far better to use basic foods as suggested above and merely augment these items with any of the wild foods that can be gathered. Often there are cases of a supply of chickweed failing just when needed most; the habit of relying on a daily supply of this food causes parent birds to search for this item if it is missing. By using the standard basic items in plenty, one can often pass over such periods of wild-food failure with no ill effects.

There is a pale-headed species in Eurasia which hybridises in the wild with the western form. This grey-headed form has a slightly smaller red 'blaze' without the depth of red which occurs in the European race. Mention is made of this inter-breeding, which occurs in the north of the former USSR, merely to explain examples of this hybrid form appearing occasionally in dealers' hands.

Northern or Siberian Goldfinch
(*Carduelis carduelis major*) ****‡

General Description
The only difference between this and the native European bird is the vast increase in size in the true northern race. The white of the face is certainly a more snow-like white and there is also a visible amount of white on the nape; also, in relaxation, they show more white on the rump, as one would find in the Hornemann, hoary or Greenland redpoll. The impression one gets is that they are far more colourful than the smaller bird, but I feel this illusion is caused by the larger expanse of each hue: the red facial mask, the yellow wing bars and dark tan-chestnut back. Certainly, when freshly imported, the tannings of the breast are lighter and the surrounding white on chest and lower belly more prominent, but, after these birds moult out on a suitable seed mixture, the tannings take on a darker shade. It is when one compares the skeletal frames of these two races that one can get a true picture of the overall increase in body bulk. The 12 cm (4¾ in) length of the native form against the 14.5 cm (5¾ in) of this race is an outstanding difference. Not all northern goldfinches are this size and, as is to be expected, the variation in size is not sudden but rather increases gradually as one progresses further north.

Habitat
As the conifer forestation and scrub land of the far north increases, so this bird's environment differs from the native goldfinch and, even in aviaries, it still prefers the types of food it would be used to in the wild. I would say without hesitation that more insect life is consumed by the northern goldfinch and they have a great preference for mealworms or other livefood at all times, not merely when rearing young.

Aviary
Once again we can breed these birds on the colony system. I would venture to say they are even more inclined towards this manner of breeding because of the scarcity of nesting sites in the north and the apparent confinement of what trees there are to areas with more scrub land between, which tends to force them to live in close proximity to one another; this is also the case with the waxwing (*Bombycilla garrulus*). The minimum size of aviary for one pair would be as recommended for the European goldfinch (page 100), with appropriate increases for further pairs or nesting hens; two or three males will certainly be able to fill the needs of double their number of females.

Comfrey and dandelion can be planted for their consumption, and elders, hops and conifer for seclusion and the provision of extra insect life as well as perching and roosting facilities. This northern bird detests the rain and damp roosting as much as any bird I know, so offer plenty of dry, draught-proof shelter from wind and driving rain. I often hear these birds singing lustily in the hardest of frosty weather but the moment dampness sets in and they feel its unpleasantness, they are off to a dry roost with just the barest minimum of visits to food supplies. This avoidance of rain does not mean any non-appreciation of bathing facilities, however, and these and fresh drinking water daily are as essential for this bird's well-being as for almost all others.

Food
This bird has a great liking for mealworms and many other forms of livefood. One still needs to offer the same mixture as that recommended for the smaller race, however, and also some kind of softfood which has been carefully introduced, well sprinkled with maw seed; baby food intended for human beings is readily taken and is of great assistance when rearing nestlings. The benefit of this latter item is that, once the birds are taking it regularly, it is a simple matter to include in the mixing further useful items of one's choice for moult, rearing or general health reasons. The acceptance of natural animal protein in the form of insects is always an asset and helps immensely when creating variety in a diet; the various foods that insects have themselves just fed upon contribute in no small way, for apart from the insect itself, there will always be some of its undigested food, be it vegetable or other smaller insect life. The wild seeds, berries and greenfood, as collected for any other finch, plus thistles in season when the heads are about to lose the seeds, should be given at all times but care should be taken to collect from clean areas where no spraying of insecticides has taken place or fouling by dogs has occurred.

Breeding
One should provide conifer nest sites, either as bunches of twigs or as baskets surrounded by a few twiggy branches. A smart site can be afforded by the old-fashioned, all-wood, roller-canary song-contest cages with the door removed or wired open to prevent accidents. Fit birds will nest readily. Giving soaked mixtures, including sunflower and teazle and plenty of insect life will almost guarantee success, certainly in getting nestlings through their first week after hatching. Since mealworms are almost always obtainable and nourishing, use them as basic supply. Aphids will be present in plenty if the aviary has been planted wisely

and these too will be fed to small nestlings. Rape plants grown on the aviary floor will be eaten down to ground level when the parent birds are collecting greenfood for their young. This and other brassicas will make the nestlings' crops look almost black. Comfrey and even transplanted dandelion from the garden will all be given and the young will benefit immensely from this addition to their diet; the provision of foods extra to an austere and monotonous seed diet cannot be too highly recommended.

Greenfinch (*Carduelis chloris*)[1*****]

General Description
Without some reference to the various mutations which a veteran breeder experimented with many years ago, the description of these birds would not be complete. There are lutino, or non-melanistic yellow, birds with pink eyes; cinnamon specimens showing merely the resulting hues of yellow and brown; silver or grey birds which illustrate only the grey pigment and a further silver or grey phase which shows yellow also, this being in the wings and tail. Now in addition to these, we have a specimen apparently illustrating grey and white; this bird is almost charcoal-grey in appearance, exhibiting no yellow and, where one would expect to find this colour, it has white feathers. Whereas the other mutations all have supposedly a proven sex-linked factor, this has yet to be conclusively proven in the last mentioned; it may well be of a recessive nature. The late A. H. Scott of Fordingbridge, Hampshire, was the pioneer in this field; he bred many birds of various mutations and I exchanged with him lutino siskins, albino blackbirds and cinnamon thrushes for lutino stock many years ago. It was just after World War 2 that he encouraged me to investigate the lutino and cinnamon mutations of the greenfinch; many of my own rare-feathered mutations of the different species came from other countries.

The normal greenfinch is a very well-known bird with a breeding range covering much of the western half of Eurasia, extending into the former USSR as far as 60°N and into Scandinavia as far as 65°N and found south as far as North Africa and from Ireland to Iran. About 15 cm (6 in) from bill tip to end of tail, it stands boldly, solidly cone-shaped. It is a good-coloured specimen, predominantly green and yellow but this varies with the area inhabited. The northern race from Scandinavia and the former USSR, of which I have kept many, all seemed a little larger though there is not the great difference in size that is found in some finches. It is, however, browner, the yellow in flights and tail still clearly shown but the body colour in both male and female being diffused with lightish brown. The best-coloured males are a yellowish-green, slightly shaded with olive, the yellow becoming accentuated on breast and rump. The wing bars, carpal joints and each side of tail are bright yellow. A female greenfinch is usually much duller in her plumage, browner and having less yellow, yet I have known some almost as bright as males.

Habitat
A bird of forest, farm, woodland, roadside, indeed garden, it keeps mainly to the lowlands. It is now common in winter and can be seen feeding from peanut-holders in gardens, placed there for the tit family during hard weather. Wherever there are thickets and tall hedges, particularly hawthorn, this bird will be found. With its strong conical bill it will eat almost any seed, or break berry stones such as that of the hawthorn fruit, in its search for food; of a hearty appetite – bordering on the greedy – it will spend hours on farmyard screenings or spilled grain. If seed is placed in a garden during winter months, it will chase others from the site to secure the choicest foods.

Aviary
This is the most accommodating of British finches, certainly the more domesticated among them and as a child I bred them in a small cage. Later, with advice from others more experienced than I, they were housed in aviaries, where a better physical specimen can be produced. An aviary 2m (6½ ft) in length and height by half that measurement in width will house a pair comfortably and provide ample space for breeding operations. On many occasions I have placed more than one female with a male; quite often this can bring good results but, depending upon the character of a bird, sometimes it brings nothing but problems.

Always keep to the rule of having a neutral feeding area when keeping more than one pair together. For everyday use, plant the floor space with brassicas and comfrey. As greenfinches are rather destructive little will grow in the enclosure apart from elder, hops and juniper, but if the aviary is enlarged other conifer growth can be included.

A minimum of shelter is required for this hardy species, and as long as it has a dry roost with the roosting perches renewed often to avoid excessive fouling and consequent scaly-leg outbreak, it will remain happy for many years.

Food
A coarse feeder, the greenfinch will live well upon screenings, with such extras as pigeon conditioning mixture, pine nuts, safflower, a little

groats, wheat, rape, linseed, hemp, niger and sunflower; this mixture should be soaked for at least 24 hours and strained of water. Offer this in sprouting condition once frosts are finished and carry on until after all adult and young birds have completed moulting and frosts are again imminent. Provide only dry seed during the winter months. Almost all my greenfinches take a variety of livefood, as does the wild specimen, and, as previously mentioned, they will avidly devour growing vegetation. Hop seeds covered in rich yellow pollen assist them when moulting and I am quite confident that their colouring improves immensely from their inclusion in the diet. Give seeding grasses, weeds of many kinds, rowan berries, hawthorn fruit, dock seed and chickweed; give also Brussels sprouts which will be appreciated when other greenfood is scarce.

Breeding

This is one of the easiest birds to propagate in captivity and there is little needed in the way of seclusion for nesting sites. A wicker basket, a few twigs of conifer to shield it and some cover from rain and direct strong sunlight, provide the hen with comfort to carry out her incubating duties and rearing. Offer mosses, animal hair, coconut matting or short lengths of teased-out coarse string, a few small wisps of cotton-wool, dry grasses, perhaps even a few twigs if a natural nest is being prepared for a fabricated site or a site in a growing bush or tree. Once a female decides she wishes to nest, it will be only a day or so before the job is completed. The clutch of eggs, as with most finches, will usually number between four and six.

Many birds will accept some form of rearing food which should of course be introduced long before the breeding season commences. They will give their nestlings much greenfood and live insects; greenfood such as cabbage and comfrey can be taken fresh from that growing in the aviary but many wild plants exist which are acceptable to seedeaters and can be provided in an effort to vary the diet. During the change-over from soaked and sprouting seed to hard seed in late autumn and vice versa in the spring, there should be a period of about 14 days when the birds receive 50 per cent of each: there ought not to be any abrupt transition from one to the other.

Pine Grosbeak (*Pinicola enucleator*)****

General Description
The largest finch of our region, about 20–21.5 cm (8–8¼ in) long, is amazingly tame. I have had these birds direct from the wild habitat of the Arctic Circle and Siberia and they have alighted upon my head and shoulders, readily accepting sunflower seeds, tugging hair from my head and showing no signs of fright even when I moved around. They are heavily built, resembling in shape the northern bullfinch and found on handling to be far larger than a starling. The body of the male is rose-red suffused with a blue-grey background shade on flank and rump. White wing bars are very conspicuous; the wings and tail are darkish but dull brown; the bill and legs are black. The female is more greenish but the rump, head and upper breast show a bronze shade instead of the male's red. Young of the year resemble the hen but have an over-all greyish-green wash. The bill at this stage is slightly paler but reaches the same large proportions, stout as in the bullfinch and ideal for taking many buds and large seeds. The plumage is soft and downy, not quite to the same extent as the waxwing (*Bombycilla garrulus*) but certainly sufficient to serve its purpose of protection against the bitter cold of their natural habitat. These finches are normally seen travelling in family parties.

Habitat
The subarctic birch woods and open pine forests of the Arctic Circle and Siberia provide this bird of northern Europe with its diet of seeds, buds, fruit and berries. Its true breeding range also takes in North America and Canada.

Aviary
Although rather like the crossbill in their habit of eating and then sitting in the same position for long periods, thus acquiring a lazy reputation, pine grosbeaks in fact have to work hard foraging for food and tend to conserve energy by sitting for a long while, slightly puffing out the feathers for insulation. They are strong fliers and require a good length to their enclosure with a minimum of 4 m (13 ft), though the ideal would be nearer 6 m (14¾ ft), and with a width of no less than 3 m (9¾ ft) and a height of about 2 m (6½ ft). Provide a small roofed section at each end as a shield from driving rain, with perhaps a little side shelter as well for added comfort. They can of course put up with intense cold, but damp roosting conditions over a period invariably prove detrimental to general well-being. They readily take to new foods, such as any of the brassicas, comfrey, sowthistle, dandelion, chickweed, dock seed, rhubarb

seed, all fruits growing wild and apple. The rowan berry is a natural food for them, as are buds of birch (*Betula*), willow (*Salix*) and the small young cones of most conifers. Bearing these foods in mind one should plant an aviary to include as many as possible. Although the planning of an aviary need never be too complicated, it should allow for the natural growth of as many basic foods as it will comfortably hold.

Food

The basic diet can be of the larger seeds, as one would use for hawfinch or crossbill: sunflower, a good-quality pigeon conditioning seed, a little each of wheat, groats, safflower, mixed millets, buckwheat, rape and niger can be soaked as usual and given strained. Fruit is essential; give apple all the time, a few soaked household currants, wild berries when available and almost any of the greenfoods collected for the more common finches. This bird seems quite willing to experiment when softfood is offered and will invariably accept a baby food which has been mixed with milk and lightly sprinkled with one of its favourite seeds.

The inclusion of fresh fir branches whenever possible will help eliminate boredom and encourage this bird to search actively for choice items. These branches must be fresh and not dry withered specimens where the sap has disappeared.

Insects are a valuable part of the daily food intake; mealworms are taken avidly, along with many other easily cultured specimens. These birds quite often take flies on the wing, just as they would take mosquito or mayfly in the wild. Insect life is essential if one is to breed from these birds. The young will not fledge without a large supply of animal protein in the diet.

Breeding

The condition necessary for these birds to breed can only be achieved by providing plenty of livefood and a good varied diet, not just prior to their usual month of May for nesting, but throughout the winter. They are slow to respond and a whole season can so easily drift by with negative results unless real efforts are made in respect of their diet and the inclusion of rich, stimulating, but non-fattening items; the extra long flight will help and, with the feeding area sited in the centre, a maximum of exercise will be taken. Provide shelf-type nesting sites, made from branches of conifers lying on a stout, horizontally erected framework securely fixed at each end, spaced about 20–25 cm (8–10 in) apart. These birds nest in similarly well-splayed branches when at liberty and only once have I ever had them nest in a large wicker nest basket, many years ago. Offer a few handfuls of twigs, mosses, rootlets, fine grasses and animal hair, such as dog-hair combings, cow hair and horse hair cut in lengths of no longer than 10 cm (6 in); do not be surprised by the numbers of twigs they expect. These are used to form a basal platform and outer walls and, even in captivity, the birds will persist in taking these vital precautions against cold reaching their nest contents. On to this base is then built the nest itself which resembles a larger version of the bullfinch nest, but is lined with grasses, moss and lichen, with rootlets and hair forming the inner lining. This nest is normally built very near the trunk of the tree on a strong branch.

The clutch of eggs rarely exceeds four; they are deep greenish-blue with dark purplish-brown blotches and spots and smaller, even darker, markings between. The female does all or most incubating whilst the male brings her food to the nest; incubation takes the usual 14 days but, as with a number of birds in the far north, this can vary by about a day either way, depending upon weather conditions at the time.

Although a remarkably tame bird (mine would take mealworms from me whilst incubating), as much privacy as one can allow at this time should help towards success. I find that some birds, even very tame ones, can feel reluctant to divulge their nesting habits and tend to sit too tightly on young, with adverse results because of the lack of feeding that occurs.

If one is fortunate enough to have young hatch – and the eggs should be fertile if the pair have been allowed an aviary to themselves – offer as many types of livefood as can be obtained. It will soon be noticed which is their favourite for rearing purposes. The large wax-moth larvae of the American species were the favourite of my birds; mealworms came second and, after they had all been eaten, the birds would accept almost anything. At this time seed will also be given to nestlings, but many buds would normally be given as well, so cut the seed intake during the first 8 days or so of the nestlings' life. The best seed at this time is sunflower, soaked and just commencing to sprout. My birds took large amounts of sowthistle and comfrey, a lot of apple, Farlene baby food to which had been added soya-bean meal, ants 'eggs', Abidec multivitamin drops, calcium lactate and milk, but they would eat it without additives, just mixed with milk.

I failed to rear the young of these birds by natural means, which entailed hand-rearing them, and have no hesitation in accepting full responsibility for this as I foolishly allowed them to remain in a large aviary holding a mixed collection of birds instead of removing them at the first signs of them carrying nest materials. This is one of the few regrets I have over my years of bird study and I cannot overstress the importance of breeding from some of these rarities whenever such opportunities arise.

Scarlet Grosbeak (*Carpodacus erythrinus*)****

General Description
Sometimes called the rosefinch, this bird is about the same size as the greenfinch (15 cm or 6 in). The adult male is brownish on the back and sides but a rich crimson about the head, breast and rump. There are two pale wing bars. The bill is conical but thick, as in most bud-eaters, and similar to the bill of the greenfinch but not so pointed. A juvenile male lacks the red in the plumage, being more grey-brown, whilst the female is even more dull, being totally brown, a pale soft colour slightly striated. There are many members to this genus in both the Old and New World.

Habitat
This finch occurs throughout most of the Palaearctic regions and seems to thrive equally in mountainous regions and lowland reaching as far north in Scandinavia and the former USSR as 68° N. On the whole, one can say that they nest more in wooded areas of willow, poplar and alder than in coniferous forests, unless the latter are rather open. Their feeding habits always remind me of the bullfinch with its taste for buds of fruit trees. Wild apple, rowan, willow and birch supply this bird with bud sustenance. The seeds taken are more in line with those of the goldfinch (dandelion, sowthistle, grasses etc.) but its choice of fruit and berries again resembles that of the bullfinch, with rowan seeming the favourite. Much insect life is also consumed but appears to be taken from trees and shrubs, rarely from the ground, and so it seems that caterpillars are the main source of animal protein. This is rather a slow and late-moulting bird, invariably retaining the old plumage until it has migrated south.

Aviary
This is an active bird but neither wild nor over-nervous when housed in an aviary. Because of the flight and constant movement, the enclosure should be longer than average, say 3 or 4 m (9¾ or 13 ft); 2 m (6½ ft) high and 1 m (3¼ ft) wide as a minimum.

Remembering this bird's fondness for buds, we must consider planting the ground with a wide range of greenfoods; brassicas, comfrey, spinach and dandelion are most suitable. Include a few conifers if possible to give a natural and attractive setting. The roof should include a slight sheltered portion at each end. Place the food in the central area in a cleared space so as not to impair the flying facilities. Like almost all northern birds, these can stand cold frosty weather but are not too happy when subjected to excessively damp conditions.

Food

Sweet apple seems a vital part of this bird's chosen diet and I have found that, if fed this food, the birds rarely if ever look sick and breed to a ripe old age. Another item is fig; open one out to show the seed content and every piece will soon be devoured. They certainly eat considerable amounts of fruit in the wild so presumably these fruits are so readily accepted because they taste like the fruits which are taken naturally, although I have only seen them eating berries and seeds when watching wild specimens. Hawthorn berries, fruit of the rowan and all soft fruits, wild or home-grown, are taken freely in season. They are given two seed mixtures. The first is soaked for 24 hours, washed and strained. We make this up from pigeon conditioning seed and, before soaking, add small pine nuts, buckwheat, safflower, sunflower, mung beans and a few currants. The second is a dry mixture composed of both British and foreign bird conditioning seed mixtures, to which we add pinhead oatmeal and ground peanuts, plus a smaller amount of teazle, chicory, and lettuce; a little of this is given daily.

Rosefinches will very quickly accept a softfood. Supply one normally readily accepted by canaries; mix it with milk to a rather stodgy paste, then liberally sprinkle it with perilla seed. The perilla will ensure that the food is tasted and consumed. Until eaten freely, just embed the seed into the surface of the softfood with a thumb. This food, once taken, is useful for introducing colour food at a later date. Mealworm consumption drops almost to nil during the winter months but will be resumed with commencement of breeding, when large numbers of these and other insects become necessary while young are being reared.

Breeding

In the aviary the vast majority of these birds have nested and reared successfully at all times, making their own nests in the higher sites offered, normally just below the roof, and in a nest basket or similar base. It can be difficult to picture a nest in the wild, since, in captivity, the capacity of the man-made base rather limits the size of the nest. Those nests which I have seen in the wild have been sited mainly in low vegetation and were rather untidy, somewhat larger than a linnet nest, and with a very straggly finish on the outer top edges. The sites chosen are similar to those of the yellowhammer: low bushes well grown with tall grasses: in captivity, despite the tendency for higher siting of the nest, low sites will not always be ignored. Try to grow a dense bush (juniper seems a good site) in the aviary with plenty of tallish grass or reeds surrounding it. If this is not practicable then fabricate one low against the side of the enclosure with another similar site at about eye level. The nesting materials will be dry grasses, rootlets, coconut fibre, teased-out string in short lengths, dog hair or similar materials. Vegetable down or hair will form the final inner lining.

The eggs are a deep bright shade of blue with a few dark squiggles and spots marking their rather striking colour.

When housed on their own these birds seem quite reliable yet, in the company of a mixed collection, they seem nervous when breeding. In all probability the other finches to be found in most aviary collections are either rivals for the nest site or rather fierce when compared with the quiet nature of this bird, so docile in its behaviour that I have seen siskins and redpolls chase it away. Our birds were naturally reared and due to an unexpectedly short period of incubation, a mere 12 days, I lost a nest of young before I realised that they had hatched, but I have since had pairs of these birds nesting again.

For the rearing of young a good regular supply of livefood is essential; mealworms will usually form the basic food but do try to provide a wide variety. Although I seldom recommend maggots, due to the ever-possible danger of botulism, they can be used if they have been kept in clean bran for a period of 5 days and the black digestive tract is cleared.

Hawfinch (*Coccothraustes coccothraustes*)****

General Description
The second largest of our finches, measuring some 18–19 cm (7–7½ in) long depending upon which area it hails from, is rather top-heavy in appearance. Its very large bill and head are set upon rather thick-looking shoulders and the body tapers conically to a short tail. The over-all impression is of a strong bill and powerful body. The colouring is predominantly orange-brown and it has blackish wings, a rich dark-brown back, white shoulder patches, chestnut head and face, and greyish nape; it has a black bib with the breast pinkish-brown, fading to white at the ventral area through a most attractive range of browns. The four inner primaries and secondaries are curiously shaped, being curved at the ends and exhibiting a small notch at each side. The bill, which is bone-pink during the winter months, gradually changes to gun-metal blue with the advance of breeding fitness. The female is paler and more of a dull brown. Young hawfinches do not acquire the black bib until their first moult. Before this, although rather resembling the hens in dullness, they have a distinct suggestion of breast bars and are yellowish in general appearance. The young can be sexed in nest feather by the amount of black around a young male's bill and his darker

underparts. Regardless of any other characteristic the bill alone will identify these birds from all other finches.

Habitat
The beech and hornbeam woods, forests and spinneys are the usual haunt of this bird and only in the far north, where conifers abound, does it have difficulty in finding such vegetation. There it makes full use of broad-leaved trees on the lower edges of the conifer forests. A shy retiring bird in the wild, it is one which can become extremely confiding in captivity once it acquires a taste for mealworms and the knowledge that the keeper is the source of supply. In the wild it is far more common than one would imagine as it does tend to avoid people. Evidence of its presence can, however, frequently be seen by lines of neatly shelled garden peas. Cherry orchards in late summer hold a fascination for the hawfinch as, with its powerful bill internally shaped to grip the cherry stone, and with extra strong muscles in head and face, it is well able to break open the hard stones. Occasionally even a suburban garden will hold a pair or a visiting family but more often it is a sudden 'tzik' which identifies them as being nearby. The nest is usually in a wood, a yew tree, or even a thick, neglected fruit tree. I have seen them high in hawthorn hedges with just the steely-blue bill and a pair of eyes watching me over the rim of a rather untidy nest. Since the breeding range of this bird covers so many countries, reaching fairly far into Scandinavia and the former USSR, there is the usual size variation between northern and southern birds; sometimes it is not so very noticeable with live birds, but is obvious when comparing skeletons.

Aviary
From personal choice I try to use an enclosure which runs lengthwise parallel with a well-used garden path; in this way the birds tend to get more used to human movement and gradually become more confiding. Aviary-bred specimens or hand-reared birds, however, have little fear of the human beings they know. The enclosure should be at least 3 m (9¾ ft) long, at least 1 m (3¼ ft) wide and 2 m (6½ ft) high. This is adequate for breeding and, since the male or female can become aggressive when nearing this period, the length does serve a very useful purpose in allowing evasive action by the less dominant bird. Providing a small-roofed shelter at each end of the aviary to cover a few conifer branches will provide a little seclusion when required. Although these birds will quite readily make nests in branches prepared to form an attractive site, I have always offered large wicker nest baskets in an effort to ensure the safety of eggs and young. I do this because I have

known free-standing nests to tilt badly when built on a branch site.

The aviary can be planted with conifers or elder; if other trees or shrubs are used, be ready for the buds to disappear at the first hungry onslaught. A background can be provided by hops on a frame or wire; these grow quite densely and, although many buds will be pruned, they will still encourage spiders and many insects for the hawfinches. The seeds will also be taken if allowed to mature. Although this bird has a fondness for garden peas, I do not advise that these are grown in the aviary as they are seldom given an opportunity to flower and develop pods. It is usually far more rewarding to keep to the usual brassica or comfrey growth.

The winter months need cause no discomfort to this species. They are hardy, strong birds and, as long as they are given a dry, draught-proof roost and fed properly, they will come through in excellent condition.

Food

Sunflower, safflower, buckwheat and groats can form the bulk of the seed intake, the first three items in equal proportions and the last in a half proportion. These should then be mixed half-and-half with a good-quality pigeon conditioning mixture and a small amount of commercially mixed finch seed. Livefood will be taken all the year round, but becomes the main item of the diet when any rearing is in progress: mealworms, locusts, wax-moth larvae, earwigs, well-cleaned maggots, the latter slightly dusted with Vionate, can prove most beneficial. Ripe apple should be available at all times and, with the arrival of garden fruit, such as currants and raspberries, these also can be added to their diet. Later, when rowan and hawthorn berries become plentiful, they will appreciate a daily supply of these foods. Rarely will a hawfinch, other than a hand-raised specimen, take a softfood but, if one can coax them to utilise such a valuable source of nutrition, then it does simplify feeding during the period when they have newly hatched chicks.

As previously mentioned, green garden peas exert a strong fascination to the hawfinch and I have known many birds to be blasted out of existence in some countries by irate gardeners. I have not advised the use of cherry stones; I find that, when these birds are supplied with enough of the above foods, they tend, in captivity, to ignore this fruit stone. Offer fruits of hornbeam, sycamore, beech and wych elm.

Breeding

Compared to other members of the finch family, the hawfinch can prove a little difficult to breed, but I would hasten to make clear that the problem can be overcome if one is willing to specialise and cater for this event. One can either culture or buy insects suitable for their use. Whichever is decided upon, they will require large amounts if they are to rear a full nest of young to maturity. A sudden change of diet to almost 100 per cent mealworm can upset the parent birds, with sometimes fatal results, so the introduction must be gradual. Maggots are safe if well cleaned, but will require some form of multivitamin additive, either dusted on them in powder form or supplied as a few drops of concentrated vitamin liquid. Whatever is intended for the birds' use should be introduced well in advance of the nesting season. The natural food of these birds when rearing is the green tortrix caterpillar found on the oak tree. I know well from personal experience just how hard it is to collect this creature, but if one does take the trouble to obtain them they will be found to be full of vegetable matter, mostly in a partly digested state. One could also offer locusts which have been fed almost entirely on grass. It is rare for hawfinches to offer nestlings any seed whatsoever until the young are at least 8–10 days old and I have, on crop examinations, found the remains of surprisingly large beetles in quite small nestlings.

The nesting site need be only a largish wicker basket lightly surrounded by conifer branches. Once breeding condition is reached the birds seem to throw all caution to the winds and will incubate in full view of their keepers. Supply a few fine twigs, plenty of coconut matting or strong bristly string teased out to individual strands and given in a moist condition; horse, cow or dog hair will be taken for the lining and a little moss is sometimes used in the wall construction. Offer a limited amount of grasses; strong rootlets will usually be used to bind this flimsy nest together.

Linnet (*Acanthis cannabina*)[1][*****][‡]

General Description
The length of this bird varies between 13 and 15 cm (5 and 6 in), depending upon whether it is from the northern or southerly area of the range. It is predominantly brown with a streaked chestnut back in the male. There are white edges in the flight and outer tail feathers and the underparts are buffish but streaked with darker brown. In the spring, the male will acquire, following a partial moult, a crimson-marked breast and forehead. The female lacks the red and is much paler brown, often greyish-brown in the head and nape, and is also just a little smaller. In the breeding season the bills will turn a dark bluish-black and the song takes on an almost continuous sound. The crimson markings of the chest will appear on males only in spacious aviaries where they have

been fed wisely on a well-mixed seed diet, wildfoods and insects and where they have been used to much flying.

Habitat
The range extends north almost to the Arctic Circle and covers most of Europe, reaching Africa and Asia Minor on the southern fringe. Although largely a bird of the common scrub lands, it also visits gardens, open parkland, railway embankments etc. Wherever one finds gorse and similar vegetation the linnet is not too far off. Low seeding weeds, such as brassicas, chickweed and persicaria, form the bulk of the diet, each one as it crops becoming the main constituent. A farmland area may be the main source of food for a few weeks and then, with the ripening of other plant seeds, open moorland.

Aviary
Despite their shyness (less so with northern birds) they are easily accommodated. A 2 m (6½ ft) by 1 m (3¼ ft) enclosure will house a pair, or even a trio – a male and two females. For winter shelter they need only the barest of roofed areas, with a little shelter from driving rain around the edges of this roofed section.

Their aviary can be safely planted with a number of shrubs and flowering vegetation. Include a few brassicas, such as rape, cabbage and kale; these will provide all necessary greenfood and later the seeds will be greatly appreciated by linnets. This bird seldom does the amount of damage to growing shrubs for which most finches are notorious. The inclusion of such natural items as gorse, broom, and perhaps hops on a rear trellis against the furthest wall, will encourage them to nest in spring.

Bathing facilities should always be available, as for all the finch family.

Food
Screenings with an extra mixture of rape seeds and a little hemp would prove ideal for most of the year, with linseed added during the winter months and the hemp increased a little during this cold spell. Screenings, however, are becoming a scarce item these days and it may well be necessary to mix one's own linnet seed mixture. For this use equal parts of canary, grass (rye), small millets, red and black rape, turnip, linseed and a small amount each of gold-of-pleasure, lettuce, maw and niger. Sweet apple will be eaten by some and ignored by others; berries are rarely taken though I have, in the far north, occasionally seen linnets eat the flesh of rowan fruit. Persicaria and all the wild-growing seeds,

such as chickweed, are avidly taken as will be the seeds of Michaelmas daisy, forget-me-not and pansy from the garden beds. Sowthistle will always be found a valuable addition to any finch diet when rearing is in progress. Linnets may not take this so much in the wild but, when kept under controlled conditions, they give it in quantity to their young. As with many of the finch family, acceptance of a softfood is an individual thing. Some readily eat such foods as Farlene (Glaxo), moistened with milk and sprinkled with maw seed and others will take a prepared finch rearing food; but there are many which seldom take such items yet eat readily all the aphids available and rear their young on these with only soaked mixtures, as mentioned above, as a supplement.

Breeding
Rarely is a nest basket or fabricated site necessary if linnets are given a naturally planted aviary. They will make do with a few branches of broom, conifer, gorse or similar growth and build a nest securely fastened in this. They invariably make a strong nest of grasses, moss, thin twigs and rootlets, lined with fine hair and wool. If providing cotton-wool do tease this out into small tufts to avoid leg entanglement. They will normally require only very small pieces of such material.

Three or four broods are often raised, depending upon the availability of suitable food; quite often if the supply of wildfood ceases then so, too, does their inclination to breed. The young of this species is noted for its wildness when first leaving the nest; even in a cage, if fostered out to a canary pair, they will run around the floor like mice. They do become tame very quickly in an aviary, provided that their parents are not over-wild and as long as they do not have to be caught up and separated.

Give as many wild seeding plants as can be gathered and make sure a good and varied soaked mixture is always available. Aphids will frequently be taken from growing vegetation such as hops or even roses in their enclosure. This is one of the first birds caged by man and the song alone makes it a very worthwhile addition to a garden aviary.

Lesser Redpoll (*Acanthis flammea disruptes*)****‡ and (*A. f. cabaret*)****‡

General Description
Vast numbers of varieties of this redpoll group exist although all are about 11.5 cm (4½ in) in length. In the lesser redpoll there are forms with a pale, almost white, breast, showing a pinkish-rose tinge, and others which are a rich rose-red from throat to deep in the chest. The variation

is due to a certain amount of natural hybridisation occurring in the wild, along with the normal tendency for southern birds to be darker than northern ones. There is red on the forehead of all redpolls although a bad moult may result in only a yellowish-gold being present; this is rare and occurs only as a result of bad management and feeding, or illness during the time of the moult. The male lesser redpoll should be a rich dark-brown-backed bird, striated upon flanks and rump and with a red forehead and chest: the small, delicate bill is of yellowish-bone, tipped with black. The female lacks the red breast, showing far more brown in this area.

Habitat

A. f. disruptes is resident in the British Isles where it often nests in orchards, in small trees and even bushes. *A. f. cabaret* is common throughout Europe in woody areas although, in some countries, it prefers hilly terrain where birch scrub and conifers abound. Thickets of alder and overgrown hawthorn are popular sites for nesting and its untidy nest, rather like that of the greenfinch at first glance, seems far too large for its very dainty and diminutive stature.

Aviary

If being bred merely for research or study of any particular phase of behaviour, these birds can be kept in a colony; they all tend to roost in one area, feed together, and eventually breed quite amicably in fairly small enclosures without undue fighting or any damage. For this reason one may house a number of pairs, or a few males with a larger number of females, in one aviary. For the best results and also to afford any student the fullest advantages of observation, give a pair or trio an aviary as recommended for the linnet. For more than this number of birds, one should increase the size of the original aviary by half for each additional pair. Only a small portion of the roof area need be sheltered, ideally the outer edges with a slight extension down the sides of the aviary, leaving the centre and greater part of the roof open to rain and sun.

Growing conifers look attractive and provide a natural setting for this bird. As it damages growing vegetation there should also be much in the way of food growing at ground level; comfrey and a variety of the brassicas will prove ideal and will tend to keep the birds' destructiveness channelled to a food subject. Hops against a rear wall of the enclosure will provide a roosting area and also food from the ripening seeds and from the aphids it encourages; this is useful when rearing is in progress.

Food

The staple diet should be that advised for the linnet (page 117), but without any of the hemp or rape. Chickweed, dandelion, sowthistle, meadowsweet, grasses of all kinds and seeds from birch, larch and alder will all be taken in season; few fruits or berries are taken but redpolls are often seen removing pips from an apple core. I have had redpolls which take softfood, but in most cases it was where these birds had been foster-reared by other finches, such as greenfinches brought up on such foods. This bird, when housed on the colony system, should be provided with a neutral feeding area with no nest sites in the immediate vicinity. One of the most pleasant ways of observing them is to hang large bunches of seeding grasses at eye level; they will then grip the fine stalks, feeding in groups and displaying their colouring and agility to their best advantage.

Breeding

This is one of the first birds that I ever bred and I am sure that most people who keep native European birds would find them one of the easiest and most entertaining to keep and breed in an aviary. Nest baskets, natural densely growing shrubs, clumps of conifer branches fixed securely in almost any sheltered site, even bunches of dry bracken, picked and bundled whilst still green the previous autumn, will be used for nesting. Offer mosses, dry fine grasses, fine twigs, rootlets and animal hair. The nest will be lined with vegetable down, fine hair and feathers, even small tufts of cotton-wool will be used but offer very little of this material and always in a well-teased-out condition as redpolls are prone to leg entanglement and similar accidents.

When the young hatch, the parent birds will take many aphids and other very small insects and, although caterpillars and wax-moth larvae will be fed to nestlings at this time, very few birds will accept mealworms. The soaked-seed mixture will be a great asset as, too, will soaked and sprouting teazle. Almost all wild seeding grasses and weeds will be eaten and the leaves turned for further supplies of insect life. The young upon leaving the nest are most attractive, smaller versions of the parents, minus the red. They often sit in a row upon a branch near the now-discarded nest making a charming sight for any bird lover.

Mealy Redpoll (*Acanthis flammea flammea*)****,
Hoary Redpoll (*A. hornemanni exilipes*)****,
Greenland Redpoll (*A. f. rostrata*)****†,
Iceland Redpoll (*A. f. islandica*)**** and
Hornemann's Redpoll (*A. h. hornemanni*)****

General Description
With the vast and subtle variations to be found within this group, one cannot be sure of the correct race unless absolutely certain of the birds' origin. On occasions I have handled birds taken in Korea, others from the former USSR, and still more from Canada and Iceland, and there have been specimens in each group which were hard to pinpoint. Measurements can be deceptive, as too can weights, and even colour variation; I have seen large brown-backed birds, for example, from the least likely of areas. Gradually, with the hybridisation which is taking place in such diverse areas as Canada and Scandinavia, identification is becoming quite difficult. Aviculture has not helped in any way, since there is a tendency to breed from two large birds or two small specimens with a total disregard of the true races involved.

I have grouped together all the large races of redpoll to prevent too much duplication in advising on their requirements. There is the hoary or Coue's redpoll from northern Europe, Siberia, North America and Canada, the Greenland redpoll from Greenland and Baffin Island, the Iceland redpoll from Iceland and Hornemann's redpoll from Greenland.

Some birds which have spent their existence in coniferous areas have slightly longer bills, adapted to their feeding habits, compared with others who feed more upon birch or with those such as the two lesser redpolls. These larger birds are in general much paler than the lesser redpoll, though much depends on the area of origin; the red of the breast area is replaced with anything from a pale, washed-out pink to deep pink and apart from this colouring the breast is almost white. The *hornemanni* races have an overall silver appearance compared with the greyish-brown of the *flammea* races; the former have fewer striations on flank and rump and both are far paler than the lesser forms. Some of the *A. hornemanni* seem almost as though they are covered with frost, they are so silver-looking. The further north the bird comes from, the more it tends to have a softer downy-like plumage. The foreheads are still rose-red in these northern varieties. In length they range from nearly 13 cm (5 in) to 15 cm (6 in) as compared with the less than 12 cm (4¾ in) of the lesser redpoll. They vary in body weight and in the general impression they give of body width and bulk.

Habitat
Those from the regions providing thick conifer forests will invariably prefer such areas in captivity and feed from the available sources of food, but others, such as *A. hornemanni* nest in conifer scrubland, which means the nest may be sited less than 1 m (3¼ ft) from the ground. During its migration to North America and Canada, and its occasional visit to northern Europe, even the latter race heads for the coniferous areas.

Aviary
In the wild these birds will have to work hard and forage for sustenance. Such a Spartan life ensures that those which survive to reach breeding condition really are fit examples. Bear this in mind when planning an aviary to house them and endeavour to provide extra length for flight exercise. A 3 or 4 m (9¾ or 13 ft) length enclosure by 1 or 2 m (3¼ or 6½ ft) wide and 2 m (6½ ft) in height would be suitable. Place the feeding tray centrally so that the distance between this and their roosting and perching areas at each end of the aviary is as great as possible. One must ensure that these birds take adequate exercise as in captivity they do tend to become lazy, eating and then sitting in one place for a longish while; this results all too frequently in obesity and deterioration of condition. Roosting areas should be sheltered and dry, one at each end of the long aviary. Plant these ends thickly with conifer, and the floor between with comfrey, dandelion and brassicas. As is often the case with northern bird species, they can be remarkably tame.

Food
I would recommend that, where obtainable, screenings are used as a base for the seed mixture. In its absence make up a mix comprising equal parts of small millets, plain canary, grass (rye), teazle and smaller amounts of maw, linseed, niger, gold-of-pleasure and lettuce. Almost all wild seeding weeds can be given, along with lettuce going to seed and near its ripened stage, seeding rhubarb, dock, sowthistle and dandelion. Bunches of seeding grasses will give hours of enjoyment to the birds and are of good food value. I would emphasise the need to use many of the various millets in aviary mixtures; they are good value for money and the millet sprays can be soaked for birds with nestlings. Few of these redpolls take mealworms unless gradually introduced to them first by having them cut into pieces on their daily seed. Their bills are not really suited for them but all redpolls are exceedingly fond of small caterpillars and soft grubs, such as wax-moth larvae; when rearing, some of these birds will feed their nestlings with large amounts of aphids. Sweet apple is occasionally taken (certainly the pips) but fruit in general is not

bothered with a great deal. A good conditioning seed mixture can be very helpful if included in the diet, as one often finds seeds in this which are never sold on their own commercially.

Breeding

These are normally late breeders, but after a few seasons in more southerly latitudes they often make a start before their normal late June. Give plenty of conifer sites; wicker baskets can be placed here and there with perhaps a little moss in the bases to give the hens the right idea. Materials can be varied slightly according to the races' original habitat but generally if one provides rootlets, mosses, fine grasses, fine twigs such as dead heather, lichen, animal hair (cow and horse hair are ideal with dog combings for the lining) and a little teased-out cotton-wool and small feathers, they will build their untidy nest, large naturally to insulate against cold. When providing the twigs, moss and rootlets, make sure they are soft and pliable by soaking overnight in water; this is wise with all the finches and though more important for the larger specimens, it will be appreciated by all as the materials are then more easily woven into position and set dry in the shape chosen.

Nestlings will need many small live insects if they are to progress as in the wild. When supplying moth larvae, sprinkle a little calcium lactate and Vionate (Squibbs) in the dish first; if one can get the parent birds to pass such useful extras on to their nestlings, the growth rate is increased, which is important during this period when basic bone structures are formed.

Long-Tailed Rosefinch (*Uragus sibiricus*)****

General Description

This most attractive species is, sadly, rare today. The body is the size of a large pale redpoll – *A. h. hornemanni* would be very near in size – and even the colouring is similar, with a silver wash over the whole bird. It has a long tail similar to that of the long-tailed tit. The forehead and lores are rich rose-red and the rest of the head is silvery-brown, just faintly tinted with rose; the back is brownish with the silver finish, again with a pinkish flush. The wings show the same hues, but the flights have whitish margins, and the rump and tail coverts are rich rosy-red. The throat is silver and this extends to the ventral area, with a rose-pink flush on the breast. The bill is bone-coloured, short but not pointed, and is in fact typical of a bud-eater. It is like a bullfinch's bill but smaller and neater. The female is similar in colour but lacks the rich colouring of the male, showing a pink wash only on the rump, upper tail coverts and flanks. The juvenile has no pink or rose until the moult and is predominantly a silver-ash colour. The tail alone measures 9 cm (3½ in), the total overall length being 15 cm (6 in).

Habitat
The open bush-covered areas adjoining mountain forests are its favourite areas. It does not appear to be very fond of the forests, remaining on the mountain slopes where a wide range of plants grows. The seeds of these form its diet, together with buds and small berries. It is found in Siberia and the Ural mountains where it roams in family groups.

Aviary
I have little doubt that the standard minimum would suffice for a pair of these birds; my own were certainly in a far larger enclosure the first time they nested but I do not feel that this is necessary. A larger aviary would, however, enable one to experiment with colony breeding. Mine were all very docile and I had about a dozen of them in all.

Food
If screenings are obtainable, use them as a base and add various other seeds to it. Hemp should be kept to a minimum. Grasses of all kinds, plain canary seed, the small brassicas, maw, gold-of-pleasure, millets, teazle, niger and many from the garden, such as marigold, forget-me-not, pansy and Michaelmas daisy, can be given. Offer all wild seeding weeds and grasses – sowthistle, probably a favourite in the wild, is taken greedily. Chickweed and persicaria are other seeds of which it seems extremely fond. Sunflower seed if soaked will be taken, as too will chopped-up mealworms from the surface of the seed mixture; many insects are eaten when visiting clusters of wild seeding plants. A rearing food such as Farlene (Glaxo), generously sprinkled with maw seed, will be taken, particularly if the rosefinch is housed with some other small bird that eats this food; the Farlene should be mixed with milk to provide extra animal protein. Quite apart from its usefulness as a food, and perhaps in rearing, it is a valuable medium for introducing a little colouring matter should this be considered necessary.

Breeding
If elder, willow and hops are growing profusely in the rosefinches' aviary, nesting sites such as those fabricated from clusters of birch twigs and conifer, or even conifer on its own, wired securely and fixed amid the growing vegetation, will encourage nesting.

Three pairs of mine nested, laid eggs and hatched young and each nest was constructed of identical materials: bents, rootlets, moss, vegetable down, animal hair, both horse and dog hair and a few small tufts of cotton-wool. The length of the tail always tends to make one forget how small the true body size is, and the nest is therefore surprisingly neat and small. The eggs are a rather bright blue with sparse small black spots on the larger end.

When rearing is in progress many aphids are taken; mealworms, when cut in small pieces and laid on their food, are eaten or quickly passed to the young. The seed consumption of the young seems rather low and the crop shows very dark with green food.

Pallas's Rosefinch (*Carpodacus roseus*)****

General Description
In fully adult males the forehead and fore-crown are such a silvery white that they seem almost to have a metallic finish. This is tinged with rose, while the upper crown, nape and back are rose-red, the back being striped or striated blackish-brown. The wings and tail are dark brown but heavily washed with red and margined with rosy-white; the lower back and rump are rose-red. The face, throat and sides of the neck, where they are rose-red, are also so closely spotted with the silvery white that the metallic sheen has to be seen to be really believed; the rest of the underparts are a deep rich rosy-red. Both sexes, when moulted well, are beautiful specimens. Certainly the female, although paler and of a softer hue all over, is none the less exceedingly attractive and more of a pastel version of the richer coloured male. Her upper parts are greyer, tinged with rose, whilst the rump is a deep rosy-pink; the young resemble the hen, never approaching the colours of an over-year male. It is a bird which acquires more red with each annual moult, provided that it receives colour food when housed under controlled aviary conditions.

Habitat
It is found in Siberia, visiting Japan and China in the winter, and frequents forests and fields. It feeds mainly upon conifer seeds, so that those forests made up of conifers see far more of them than the broad-leaved areas.

Aviary
Provide an enclosure about 2 m (6½ ft) in both length and height, the minimum width being 1.5 m (4⅞ ft). This is slightly larger than the

minimum-sized aviary in order to accommodate the male display and nuptial flight; this can be quite hectic, the male chasing continually back and forth, with the female often going to ground where the dance takes place. The dance is performed with the crest held erect, wings lowered and quivering, accompanied by a rather coarse song throughout. Provide fir branches if possible, but secure them at each end. To encourage as much exercise as possible, the food should be sited well away from any perching facilities. Maximum exercise is of great importance to some of our larger finches, such as the pine grosbeak, because some tend to do little other than eat, perhaps a little too well, and then spend ages on a favourite perch digesting. This is common when a bird which has to work very hard to find adequate food in the wild finds everything it needs more or less at its 'finger-tips' in an aviary. Offer bathing facilities at all times.

Food
Soak a mixture of pigeon conditioning seed, small pine nuts, safflower and a mere sprinkle of sunflower and mung beans and leave in water for 24 hours; rinse well and strain, then serve. Offer also, but in another dish, a dry mixture of both British and foreign bird conditioning seed, together with fresh ground peanuts, pinhead oatmeal and smaller quantities of teazle, chicory and lettuce. Give sweet apple at all times and dry figs split open to disclose the seeds. Berries of all kinds will usually be taken: wild types such as rowan and hawthorn, and blackberry too. Ripe rhubarb seed and dock, when both are reddish-brown are eaten and seeding dandelion heads, sowthistle, chickweed etc. should also be given. In fact, offer everything that would be given to bullfinches. If a bush of japonica grows in the garden, do not waste the small quince-like fruit, which are absolutely packed with pips; all seedeaters are very fond of these, despite the fruit flesh being useless for birds. Softfood will be taken when introduced carefully; as previously mentioned (page 19), make it rather stodgy, cover it with perilla seed, and they will soon eat it all.

Livefood, although rarely taken in great quantities during the winter months, can be greatly increased with the arrival of spring. When young are being reared, for the first 8 days at least, considerable numbers of mealworms and wax-moth larvae will be eaten and passed to the nestlings. Also at that time, from the sixth day onwards, the young will require far more sunflower to be soaked with their mixture.

Breeding
These birds certainly seem to have a built-in preference for conifers, not only for perching but also as nest sites. They will readily accept almost

any nesting receptacle, whether a basket or wire-mesh base, but again they do show a particular liking for small conifer twigs to surround their nest. It is not due to shyness for, when incubating, they become exceedingly tame, but, just for a short spell when building their nest, what seems to be a natural shyness over their planned site makes itself felt. If encouraged to build their own nests from scratch in a bush or man-assembled bunch of branches, the nest will be a rather loose structure. They appreciate small twigs of birch or heather, rootlets, bents, moss, fine grass, animal hair from cows or horses and even dog combings for the lining. When using horse hair, do make certain that it is cut into lengths not exceeding 5 cm (2 in). The eggs are a bright rich shade of blue, heavily spotted and slightly glossy in appearance. The female alone incubates and is quite content to let her mate sing his rather continuous serenade from a nearby perch. The rather coarse song, lacking in sweet notes, becomes monotonous long before hatching is due. Regardless of his failing where song is concerned, and his clumsy earlier advances, he is an attentive mate, feeding the hen at the nest right up until the young are about 6 days old. When I moved to East Anglia in 1980, not so very long before the birds were due to commence breeding, with new breeding pens to be constructed and having little in the way of leisure hours to spare, I wondered how long it might take for all our birds to settle in. I need not have been concerned over the Pallas's rosefinches, for they were one of the first species to commence nesting; they reared successfully in 1981.

Serin (*Serinus serinus*)****

General Description
For those who are familiar with the border variety of canary I would say that the serin resembles a small chubby version of this, about 11.5 cm (4½ in) long from bill to tail. If one visualises a green border canary then thinks of darker striations on it, one has an almost perfect picture of the serin. Predominantly green with a small neat bill, the male has forehead, breast, rump, nape and eye stripe of yellow; the female's colouring is more dull with far less yellow in the plumage. The young can be duller still, being slightly brown-green with even more striations than the female. The song is a very sweet little melody with perhaps not much variation, but so continuous and happy-sounding that I always find it a pleasure to hear. It is normally a lively little bird, seemingly fit and happy in warmer dry weather.

Habitat
Gradually becoming more common throughout Europe, this bird now reaches Sweden in its northerly range and breeds in Poland and the former USSR. It is a woodland bird by choice, but where this ends and cultivation begins, their nests can be found in vineyards, olive groves, garden hedges and parks. Although becoming widespread, it is still very much a Mediterranean bird.

Aviary
The minimum-sized aviary is ideal for a pair of these birds and, even with this size, two females can be included with the one male if so wished. I have never yet known any real aggression show itself with these birds. Give plenty of winter shelter; although I would not recommend artificial warmth, they are better for having a small shed-type shelter attached to their outdoor aviary. In this manner they may take advantage of sunny winter days yet remain out of the damp when necessary. Construct this shelter so that plenty of natural light enters and clearly illuminates the interior. Where possible place the food inside. I would not advise shutting these birds inside for the duration of the winter as they are early nesters, often undertaking nesting operations in March, and if shut indoors for all the winter their condition is not at its hardiest just when it needs to be. If it is not possible to build a shelter, then ensure a dry draught-proof roost.

The aviary can be well planted with conifers with plenty of ground cover of food value: dandelion, brassicas and rye grass will provide a natural diet.

Food
A seed mixture should contain plain canary seed, mixed small millets, grasses, lettuce and red and black rape. Very little niger or linseed need be included but a canary conditioning seed should assist in providing some of the small seeds so enjoyed by this bird. Do endeavour to obtain screenings for this bird, thus giving it a free choice of wild seeds. Dandelion, chickweed, shepherd's purse, persicaria and clusters of seeding grasses suspended from the aviary roof can be continually pecked over and the fallen seeds gathered from the floor.

Softfoods will be taken if gently introduced. One of the favourites for rearing has been Farlene (Glaxo) baby food, mixed with milk; in the early stages of rearing one can always mix extra calcium lactate in such a food to improve the bird's bone structure and forestall any danger of rickets.

Insect life will be given to nestlings, if available, in the nature of aphids and such-like. I have had one or two birds take maggot chrysalids, opening one end and drawing out the contents as goldfinches do, but this is rare in serins.

Breeding
This bird can be kept in a manner almost identical to the siskin. Small wicker nest baskets will be accepted and used readily. Natural nests may also be made in almost any growing vegetation or suitably fabricated site. Preference is always shown for conifer sites. Materials should include mosses, fine grasses, lichens and rootlets; the birds will require above-average amounts of hair and feathers for lining this structure and the whole thing is a very warm little nest; in the wild, serins' nests are very well concealed.

Soaked seed, wildfoods and aphids all contribute towards successful rearing and the young, attractive little creatures, are well worth any effort.

Wild Canary (*Serinus canarius*)****

General Description
Apart from being a little larger than *Serinus serinus* and showing more grey on the back and a little less yellow elsewhere, these birds are very similar. In an aviary they behave in much the same way. The surprising thing is how man has managed to develop, from this bird, the domesticated canaries of today. The length is 12.5 cm (5 in).

Habitat
The Canary Islands, Azores and Madeira form the home of this rather dull little bird. It frequents woods, orchards and gardens where the song can be heard from the perch and as it flies over-head.

Aviary
Provide a similar structure to that advised (page 128) for this bird's close relative, *Serinus serinus* and treat in a like manner in all ways.

Food
Feed as for *Serinus serinus* (page 128) or the domesticated canary but tend to offer more grass seed than for the latter.

Breeding
I may seem a little uninspired by this bird but so much has been written on the domesticated canary that, when a wildling of the species does

occur in captivity, one can add only a minimal amount of advice. I would merely suggest that one treats a freshly taken specimen not as a canary but as a wild bird, in particular *Serinus serinus*. These birds will nest quite readily once acclimatised and used to aviary life. They adapt to their surroundings quickly and soon begin to rear successfully. However, in the early stage, they should be treated as wildlings and be given all assistance in rearing, with the provision of wildfoods and grasses. Although a canary by name, they are in need of specialised treatment at first, such as one always gives a newly taken bird.

Red-fronted Serin (*Serinus pusillus*)****

General Description
The male is of a striking appearance with the fiery-red forehead contrasting with the black of the nape, side of face and breast; the chest fades to yellowish-green on the lower belly and the wings, although mainly blackish, are streaked with a tawny-rufous shade which extends to the tail. The ventral area is yellowish-white; the wings have yellow patches and there are also yellow feathers at the side of the tail. A female is duller and lacks the red forehead. The juvenile of this species is paler on the lower belly than the parents and very dull and blackish around the head, nape, back and breast, showing merely a little of a yellowish shade on wing butts and chest.

Habitat
A high-altitude bird of coniferous forest, this is a species I have kept on a number of occasions but never studied in the wild. Mine came from the former USSR but its usual habitat is the high mountainous areas bordering the former USSR and Turkey.

Aviary
When siting the enclosure for these birds, give a little thought to their natural environment, altitude in particular, and erect the aviary in a cool site, one where perhaps the sun only filters through to the enclosure. I recommend this because, when my own birds have nested and sun has struck the enclosure, the hens have looked most uncomfortable, raising themselves from their nests with bills open. Although there is certainly no scarcity of sun in their natural habitat, the high altitude does provide very pure thin air. Living lower, at sea level or thereabouts, they may not feel at their best. A dry roosting area is essential if they are to have a normal life-span. They seem to detest dampness although they are

frequent bathers. No heating is required in the winter roosting shelter.

The minimum-size enclosure should be in the region of 2 m^2 (6½ ft^2) with a similar height. They are very active fliers, circling and diving at high speeds, so one should give ample space for their manoeuvres. Conifers can form the background provided that they do not cause obstruction; they should be kept to corner sites and perimeter. The shelter need not be a very large one, merely enough for roosting and avoiding excessive dampness.

Food
Here we have a bird which, in the wild, eats large quantities of seeding grasses and other typical serin foods. Feed as you would *Serinus serinus* (page 128). Do not give too much niger as this bird very quickly becomes addicted and will make it its sole diet if allowed to. Small insect life is taken and, when rearing, there is a continual search for aphids and other insect pests among growing vegetation; with this in mind it is advisable to plant hops along one side of their aviary. They seem to take dandelion, sowthistle and seeding brassicas in large amounts. Hemp should be omitted from their diet. There are many wild foods to select from, including the numerous seeding grasses which form their natural wild diet. The use of a commercial wild-seed mixture or screenings will greatly assist their general health.

Breeding
These birds take to wicker baskets or conifer sites with little concern for nearness of human beings. None of my birds has ever seemed at all shy at this time. One can leave them to build a nest in conifer shelves or growing vegetation once the materials have been provided: mosses, rootlets, dry grasses, lichens, hair and feathers should be made available.

Give soaked seeds at all times after the frosts of the winter have ceased right through the spring, summer and autumn, until after completion of the moulting season. Then, over a period of a week or two, hard seed should be resumed mixed half-and-half with the soaked.

Siskin (*Carduelis spinus*)[****‡]

General Description
This bird is often considered to be one of the smallest of the finch family but it does vary in size depending on its region of origin. As usual, those from the far north are longer and also slightly bigger in body bulk. The length ranges from 11 to 12 cm (4¼ to 4¾in). The variation is not

noticeable in the field and, indeed, the larger of the specimens are found as residents only on the bird's very northern limits. An adult male is an all-over green and very yellowish around the face, rump and breast, with large yellow wing bars and yellow also at the sides of the tail. The crown should be black, darker and more glossy with age and there is also often, but not always, a black bib just below the bill; this varies in size and depth of colour from one male to another. The back is striated with black; wings and tail also show dark almost-black feathers. The lower belly is unstreaked, with the yellow fading out to white in the ventral area. The hen is paler, more greyish-green and lacking the black on the crown and the bib marking; the wings also are greyish-green but do show the wing bar, although in a paler shade of yellow; the whole body is more streaked except for lower belly to vent. A juvenile resembles the female somewhat but is even more streaked. With their first moult the young males will acquire a pale black-flecked crown.

Habitat
They reach almost to the Arctic Circle in Scandinavia and the former USSR. Some are resident all the year round in their breeding grounds though the vast majority spread during winter over the whole of Europe. The conifer forests are the favourite haunts, a great deal of the food being taken from trees. These same conifers provide the sites for their small compact nests, which are usually constructed on a high horizontal branch.

Aviary
Here again we have a small finch which adapts well to colony breeding. A surplus of hens should always be present where this method is operated. The enclosure for one pair will need to be 2 m (6½ ft) long by the same height and 1 m (3¼ ft) wide. Perches should be provided by growing conifers at each end and the feeding area must be situated centrally. The ground should be densely planted with comfrey, brassicas and dandelion. A small sheltered portion at each end should cover a few bunches of heather and this roofed section should extend downwards a little at both ends to provide an ample and dry roosting area. Keeping the main perching areas well apart to enforce exercise is beneficial to a bird such as this, which tends to overeat oily, fattening foods. The deterioration in general condition which occurs as a result of too little exercise is reflected in fewer hens nesting and rearing successfully. An aviary double the above recommended dimensions would house up to three pairs and, provided that the inmates are fed wisely, breeding attempts ought to be successful.

Food

Very often I have fed siskins during the non-breeding season with seed mixture left-over from other birds, such as the goldfinch. Any signs of obesity have thus been avoided. When breeding and rearing are in progress I increase the mixture to include soaked teazle, small sunflower, extra millets and niger. The ideal basis for their seed mixture is screenings or a commercially-sold wild seed but, if these are unavailable, one can mix the smaller millets, rye grass, plain canary and teazle with half amounts of linseed, maw, gold-of-pleasure, lettuce and, if it can be collected during the winter months, dock seed. These will add up to a varied and nutritious standard diet. As for all finches, I would advise soaking the seed for at least 24 hours and straining it for a short period before offering it to the birds. It is vital that one feeds in such a manner until all the birds have completed their moult. Once having started this method of feeding, there should be no changing in mid-season. Where space permits, I would advise the inclusion of thick hop growth at the rear of the siskins' aviary; apart from the enormous quantities of aphids this will supply, the birds enjoy roosting in this plant. When collecting wild foods, remember that, in the wild, siskins will eat seeds from trees such as larch, pine, spruce, alder, elm and birch. The favourite foods at ground level are dandelion, chickweed, sowthistle, persicaria, meadowsweet, fat-hen and dock.

Breeding

Almost all siskin hens will build a natural nest in conifer branches. It is ideal to have such vegetation of the right height growing in their enclosure. Until these are established, however, and the top growth pruned to thicken the lower growth, it may be necessary to provide and fix conifer branches in the required nesting-site areas. Nest baskets should also be fixed in position amid these branches, fairly high but sheltered from heavy rainfall. Once these baskets are established, it matters little if the conifer turns brown and drops needles; the aviary need not be entered again. Branches alone will soon turn brown and become unsightly, and they are of little use for nesting in. The nest itself is a neat cosy little thing, well woven together and comprising mosses, lichens, very fine rootlets and grasses; it is thickly lined with animal hair and feathers. The resulting young make an attractive sight when fledging, perching in a row. They are diminutive creatures, frequently leaving the nest at 14 days or shortly afterwards.

Aviary-bred strains of siskin usually take some softfood when rearing and many will eat it all the year round. Try them on Farlene (Glaxo) moistened with milk and sprinkled with maw and a little niger; once birds are using such foods for rearing there is seldom a case of rickets or undernourishment among newly fledged young. During the breeding season, when pairs are feeding young, provide as many of aphids, gentle chrysalids, wax-moth larvae or even cut-up mealworms as possible; the latter should be placed on top of the daily seed mixture.

Twite (*Acanthis flavirostris*)****

General Description
This is a brown bird, very similar to the linnet at first glance and, although the same length, a little over 13 cm (5 in), gives the impression of being longer because it is slightly slimmer. It is easily discernible at close quarters by the orange-biscuit throat and long pronounced forked tail. The male has a rosy-pink rump and a bill which is yellow during summer and greyish in winter; the white in the wing is not quite so clearly defined as in a male linnet. The overall appearance is of warm brown streaked with darker brown, lighter on the belly. The female is more striated, particularly on the chest, and the young even more so on throat and breast. They are also greyer on the head and nape.

Habitat
Considered the northern counterpart of the linnet, this bird inhabits the wild barren tundra areas and mountainous regions of the far north. However, it does breed in lesser numbers in northern parts of the British Isles, in hilly moorland terrain among heathers.

Aviary
Provide for these birds as for the linnet (page 117). They are not quite so nervous as their close relative but require an aviary of about the same size with a conifer growing at each end. This shows them to their best advantage. Clusters of heather arranged from ground level to roof will cater for their roosting and nesting needs. Winter shelter can be of the barest minimum, with a simple roof and vertical wind- and rain-break. The wild seeds of moor and cultivated land provide much of the food required and, with this in mind, one should grow brassicas and dandelion in the aviary.

Food
Feed exactly as for the linnet (page 117). A coarse feeder because of its environment, this bird will eat almost any wild seeding weed or herb. The seed mixture, if given as recommended, will keep them in excellent

condition for many years. Rarely have I known twites other than hand-reared specimens to eat softfoods, but I have seen them take aphids and other insect life.

Breeding
In an aviary this bird will often nest in a high site, whereas, in the wild, it nests at ground level in heather and rock crevices. The clusters of heather as advised for its roosting facilities will also provide ideal nesting sites; twites will use wicker baskets but all the natural nests I have seen in aviaries were strong structures not requiring the support of a basket. The materials to offer are fine twigs – heather is perfect for this – or even birch or beech twigs, grasses, moss, all kinds of animal hair and small tufts of wool, and finally feathers for the warm lining.

This bird will definitely use small soft insects when rearing but the bulk of the nestling food intake is soft, pulpy ripe seeds. When it is not possible to collect such foods, soaked seeds including rape, charlock, radish and turnip can be added to the normal mixture.

8

Blackcap, Chiffchaff, Firecrest, Goldcrest, Warblers and Whitethroats

IF I APPEAR TO GENERALISE on the subject of warblers it is with good reason as their needs are very similar although their habitats are widely dispersed and variable. As a general guide they all tend to require slightly more fruit and vegetable matter in their diet than most softbills. They also need bathing facilities more than any other birds.

The basic food is as described for insectivorous species in Chapter 3. If a proprietary food is used, insist on having an analysis of its contents and you will be able to fill any gaps with fresh food of your own choosing.

It can be most helpful to use a baby food as a base for a supplementary diet, to which one can add a variety of items which must be included in a bird's daily intake: multivitamins, calcium lactate, colouring matter, extra protein of a particular kind. Experimental additions to the diet can all be mixed in such a food which itself is moistened with milk. A few species are slow in accepting further such food when already eating a man-made preparation, but natural inquisitiveness usually prevails and a tentative taste will normally ensure its future acceptance along with that of any additions one wishes to make. Such additives as liquid multivitamins could always be included in the drinking water, but there is a great wastage when drinking water is changed daily. One must consider as well the small amount of water taken during 24 hours and the subsequent minute dose of the vitaminised product ingested in this manner.

Ripe fruit, such as apple, pear, grapes, peach, avocado pear, date, banana and fig, should all be included in a warbler diet, as can soaked household currants, or even fresh or defrosted blackcurrants, blackberries and elderberries. Livefood can consist of wax-moth larvae or wood-ant larvae ('eggs'), the latter in either a fresh state or defrosted. A few cut-up mealworms every other day, and locusts, shortly after hatching, will be taken by almost all warblers.

Where one uses gentles, ascertain before use that the digestive tract, the thin black line to be seen running through the insect, has vanished. This is a safeguard against passing on botulism to one's stock; 5 days in fresh bran will normally clear any matter from this tract. One could obviate this by feeding the gentles on a mixture of Farlene (Glaxo) mixed with milk. One would thus be sure that the contents of the gentles would be harmless to stock, but one would have to contend with the problem of wet maggots climbing out of the container and escaping. The value of these gentles, or fishing maggots, since this is the prime purpose in breeding them, can be greatly increased by powdering them lightly with such products as Vionate (Squibbs), Casilan, Complan (Glaxo) or a few drops of Abidec (Parke, Davis).

Fresh greenfoods and root crops are also needed in the daily diet: grated carrot, finely chopped or minced watercress, whole dandelion (flower, leaves and root), spinach, comfrey or any brassica minced finely and mixed in the basic softfood can provide day-to-day variety.

Minced raw or steamed fish, or crustaceans such as prawns and shrimps, can be given in small quantities to the birds' advantage. Any of these may be mixed with the basic insectivorous mixture.

A separate food dish containing finely ground peanuts, a little perilla seed and some pigeon conditioning mixture, all served as a powder, will be found most acceptable to all the birds. A considerable number of softbills take this daily: species as small as wrens and goldcrests, not just the larger insectivores, or warblers etc., will take a fair amount.

Such a wide variety of foods as this, taken quite freely, illustrates just how varied their food intake can be when they are allowed complete freedom of choice. Trace elements present in the foods should benefit the birds.

Just prior to migration almost all warblers gain a great deal of body weight and, under controlled conditions, this persists even if they are being fed correctly. In aviaries at their peak weight many such birds go off their food when no long migratory flight is undertaken and they gradually slim back to normal. The use of plenty of fruit and vegetable matter in their diet at this time will aid this natural process. They need not be placed in the winter quarters at this time unless extremely cold weather is being experienced or forecast.

The inclusion of high protein or valuable food matter in the actual culturing (at least during the last few days) of any livefood intended for one's stock always assists in providing variety of diet. The little extra contained in each digestive tract is well worth the trouble. I am very much against providing livestock with insects which have been living on their own fat for a week or so prior to their use, with little or no additives. Every insect taken by a bird has certain foods in its digestive system and all these partly digested items of vegetable, cereal or animal-protein matter provide, over a period of feeding in this way, a greater variety than man-prepared mixtures.

Blackcap (*Sylvia atricapilla*)****

General Description
The male has a black crown extending to just about half-way through the eyes but not below; the back and wings are dark brown, the cheeks are greyish and the breast is of pale silver-grey fading to white on the lower belly and ventral area. A female is slightly duller and a little browner, with a brown instead of black crown. Juveniles are rather like the female but young males achieve a brownish-black crown. It is 14 cm (5½ in) long when fully grown. It is noted for its song, which is reminiscent of the garden warbler which it has a habit of mimicking.

Habitat
A migrant, this is a forest and open woodland species, preferring spinneys and areas of sparse trees with overgrown brambles and thickets of scrubland.

Aviary
The standard minimum size, as previously advised (page 77) will suffice. It should be planted with hops at the rear, trained on the wire, which will provide extra livefood. A thick bush such as juniper or bramble will be required for nesting purposes; site this towards the rear and under a wire roof. The aviary needs a shelter attached, which should be a shed-type of building, not necessarily heated but closed all round with plenty of natural light entering; these details apply in a temperate zone and must be varied according to one's climatic conditions. Free entry to the shelter and outdoor flight may be allowed all year round.

Food
Use the basic food mixture, with the addition of fruit at all times, livefood of a limited quantity daily, and attempt to provide a varied diet; include, as an extra, milk-moistened Farlene (Glaxo).

Breeding
A pair may be housed together at all times throughout the year but allow one pair only to an aviary. Increase livefood supplies with the nearing of the breeding season. Supply plenty of grasses, varying from thick and straw-like to fine, rootlets and all kinds of animal hair for lining the nest. Expect this to be constructed in the thick low bush provided for this purpose. The pair will need large supplies of live insects for rearing and, in later stages, will give the young softfood and fruit. Insects will form the bulk of the diet until fledging.

Chiffchaff *(Phylloscopus collybita)* †*****

General Description
This little leaf warbler looks remarkably like the willow warbler but the dark legs and the monotonous 'chiff chaff' song identify it immediately. It is, too, perhaps a little slimmer in the body, with olive-green plumage, yellowish below and with a distinct yellow eye stripe. There is a Siberian race which is far more grey but alike in other ways.

Habitat
The favoured living area of this migrant is tall woodland where it will take insects from the trees, on the wing and from the ground or low vegetation. Most of the food, however, is gathered from the leaves of trees, where it will be seen flitting daintily. It is not so dependent upon bushes as the willow warbler, except for nesting when it builds its dome-roofed nest in low, thick, bramble-like undergrowth.

Aviary
Treat exactly as for the willow warbler (page 163). Pairs can be kept together all the time but there should be no more than one pair to an aviary. Plant the aviary with a dense bramble over about 1 m^2 ($3\frac{1}{4} \text{ ft}^2$) of the floor area. This needs to be quite thick and well established, otherwise it is better to fabricate a near-natural site for their nesting requirements.

Food
Feed on the same diet as the willow warbler (page 163) but allow a little more livefood; this can be cut-up mealworms, only two or three a day, or wax-moth larvae, or thawed-out or fresh ant larvae. Most livefoods can be given in great quantity but then one has the difficulty of a semi-meating-off period when brought indoors for wintering. It is for this reason that I ensure that softfood is eaten daily.

Breeding
A low thicket of bramble, or hawthorn overgrown with bramble and tall grasses growing up through this in a tangled mass, will appeal to the chiffchaff as a nest site, but do leave this well alone; do not peep inquisitively when it is feeding. Give materials such as grasses, fine twigs, rootlets, mosses, dead leaves, hair and feathers. With the arrival of young one must remember that the parents will offer them only insects. Occasionally a hand-reared specimen does offer nestlings whatever it was reared upon itself, if that food has been in daily use ever since.

Firecrest (*Regulus ignicapillus*)†

General Description
This is very like the goldcrest but has black-and-white eye stripes, a crown more red than orange in the male and a bronze neck patch. When her feathers are parted the female has a far paler crown. The length is 9 cm (3½ in).

Habitat
It prefers low bushes, ivy-covered low walls, broom, gorse and vineyards. It does not seem to favour any particular tree but likes low vegetation and, unlike the goldcrest, has no special preference for conifers.

Aviary
As for goldcrest (below).

Food
As for goldcrest (below).

Breeding
As for goldcrest (below).

Goldcrest (*Regulus regulus*)†*

General Description
This and the firecrest are the smallest birds common to Europe, at just 9 cm (3½ in) long. The goldcrest is dull green, the male being just a shade greener than the female; this colour fades to a creamy-buff on the underparts. Both sexes have two white wing bars. The bill is a delicate one, ideally suited for taking very small insects in the wild. The crown

has no real crest but the male has a brilliant orange line, centrally from the upper forehead almost to the nape, which is edged with black; in display the male will spread the black edging feathers sideways to reveal the vivid orange colour to such an extent that it almost covers the crown. There is no conspicuous raising of any crest. The female's crown is only palish-yellow.

Habitat

Woodland areas are favoured, with a preference being shown for conifers and especially for large gardens and churchyards containing yew trees. In the winter months this bird will often enter suburban gardens where it will seek insect life in dense low foliage, bracken and ferns.

Aviary

The standard recommended minimum sized aviary is ideally suited to a pair of these birds. It should have a frost-free shelter attached to it, in which the food is placed every time. Extra feeding time during the winter promotes longevity and entices small moths and other insects into the enclosure. A yew tree at each end should be kept trimmed to well below the level of the wire roof. This encourages the birds to enter the shelter in search of a higher roosting site, and this will show them off in a natural setting. A small boggy area near a very shallow pond, well planted with ferns, should cover the majority of the aviary floor. Wild bracken, however, grows too high and prevents any sighting of the inmates most of the time.

Food

This will depend entirely upon whether the birds were taken under licence as adults, hand-reared from the nestling stage or taken as fledged young. In the case of the first, a certain period of 'meating off' will be necessary; patience and leisure time are essential to complete this operation successfully. It will be necessary to have a finely ground insectivorous food, and finely grated cheese, with a few cut-up mealworms and gentles on top. This should be placed high near a perch where it will be obvious to the freshly acquired stock. Prior to the actual taking of such birds one must have prepared other feed too; one should have a culture of fruit flies breeding near an electric light bulb in the intended shelter and also a good supply of wax-moths and their larvae. The new birds should be placed in the shed portion of their quarters and the cut-up mealworms and gentles renewed five or six times a day; their water, placed in a very shallow dish, should contain the recommended amount of vitamin additive. The moths can be liberated, about 20 to a pair of birds each day, and the larvae should be placed on the food. Wood-ant larvae are very suitable for goldcrests. Do not over-feed these as the general aim is to encourage the birds to take the insects from the top of the man-made food, acquiring a taste initially for the cheese and later for the basic food. This 'meating off' of adult birds can take a week or two and it is for this reason that I would advise would-be keepers of goldcrests to hand-rear their stock from about 8 days of age; freshly fledged young will go on to an insectivorous mixture sooner than adult birds. Where one is allowed by licence to take them, nestlings will prove far simpler to rear, since they will learn automatically to take the offered softfood (see Chapter 4).

Breeding

I do not know of anyone who has ever reared a full clutch to maturity under controlled conditions; the task of breeding at all from birds taken as adults is very difficult. Young hand-reared specimens, well used to the diet provided by their keeper, will more readily accept a basic livefood supply of moth and moth larvae. Remember that one should offer only items of livefood which can be continually supplied; One cannot hope to rear birds if changing half-way to another source of food and parents treated in this manner will often continue to search for the insects they are used to receiving.

The nest of the goldcrest is a work of art; normally suspended beneath a horizontal branch and the size of a well-developed orange, it has a small entrance in the upper half of the semi-domed structure. The materials needed to form the main structure are moss, spiders' webs and lichen; spiders' webs, indeed, are taken in such quantity that when a partly built frame was dissected for inspection, the webs were still tacky and, along with moss, formed the bulk of the material. Some goldcrests use animal hair and wool, as well as feathers, for lining whilst others use only the last. I think this depends to a great extent on what is available in their habitat.

All the small garden pests, such as greenfly, and small spiders as well as the moths and moth larvae will become vital necessities if one is fortunate enough to have young bred in captivity. If one plans well in advance, the wax-moth larvae can be in almost unlimited supply when wanted most.

Barred Warbler (*Sylvia nisoria*)[1****]

General Description
One of the largest of the European warblers, this strikingly attractive bird is very shy in the wild. It is bulky of body and long-tailed. The main colouring of mantle and upper body is pale brownish-grey. The underparts are darker grey with heavy barring from throat to belly. There is very little difference between the sexes though the female is slightly less barred beneath. A female I kept always had grey irises, whereas the male's were all bright yellow. This is an accurate sexing method. The length is 15 cm (6 in).

Habitat
Thickets, scrub and open woodland with dense low undergrowth are all typical residential and breeding areas of this migrant bird. It is rarely seen in the open; one may hear merely a warning cry and catch sight of it disappearing into the nearest low bush. The song is a melodious warbling, very attractive to the ear and repeated in short bursts, often interspersed with a loud 'charr-charr-charr' – quite a jarring call.

Aviary
The standard minimum-sized aviary can be used for this species but I have always given far larger quarters during the breeding season. These birds will need a warm winter flight indoors which can either adjoin their enclosure or be separate. I have allowed them free access in all weathers but have made certain that they fed and roosted inside during the winter.

Unless hand-reared specimens are kept, there will be much skulking in the undergrowth which is typical of the wild bird. However, once a rapport has been established between bird and keeper, they become very trusting. With their natural habits in mind, it is well to plant fairly dense undergrowth of raspberry canes, hawthorn and bracken over half of the enclosure area, leaving the rest clear and weed free. The use of some flowering ground cover can help and add to the aviary's attractiveness. My own aviaries have often held hops covering one or more of the wire sides. Quite apart from the amount of seclusion this offers, it has the usual benefit of providing an abundance of insect life.

Food
I would always advocate a high-protein diet for these birds, but they must, too, have large amounts of fruit. This can be in the form of wild or cultivated berries according to the seasons, but sweet apple should

always be available, with pear being given once or twice weekly. Soaked household currants will be greedily taken when berries are unobtainable; chopped figs and dates, if diced small, can be added to their basic food, as can most brassicas, dandelion and spinach, if well minced or chopped fine. Do not forget the daily supplies of grated cheese on their softfood. Where livefood is concerned, most species of the usual insects given to birds will be accepted, young locusts, cut-up mealworms and wax-moth larvae being favourites, together with wood-ant larvae when available.

Breeding

The nest is rather like a larger version of the whitethroats' nest, made of grasses and bents and lined with hair. It will usually be built in a dense part of the hedge well hidden by undergrowth but I have had them build in a large square biscuit tin, turned onto its side, into which had been pushed bunches of heather and bracken.

With the arrival of young, the usual large quantities of insect life must be supplied. Little fruit, if any, is given to very young nestlings but all manner of soft-skinned insects will be offered; even quite large locusts are smashed to bits, wings removed and frighteningly large pieces rammed down the nestlings' throats. When we have hand-reared these birds, almost from the egg, or where parent specimens have been feeding young in the nest, we have used, as the basic livefood, wax-moth larvae and small daily supplies of wood-ant larvae. One must be careful to ensure that a continuing supply of wood-ant larvae will be available; I have found the birds become so addicted to them that they search continually for them if they abruptly go missing from their everyday diet. It is wise to store them in the fridge or freezer and give a small daily ration. Odd as it may seem, these birds eat more of their man-made insectivorous mixture if it contains steamed fish or fish roe, yet they still like to remove the grated cheese piece by piece before taking the food in bulk.

My first barred warblers, hand-reared, were from Italy. My father, many years before, had spoken of their singing and mimicking qualities and so I decided to collect them; the less said about the cost the better, but for many weeks I speedily turned all thoughts of finances from my mind. No picture I have ever seen really does these birds justice. We had three nests from one pair in one year; they were found at the end of the season, within inches of each other, low in a hawthorn. Egg shells in the enclosure indicated that hatching had occurred, but we did not rear any young. This happened for 2 years and I was rather over-stocked at the time of the next breeding season. I discussed with old friends, both leading ornithologists, what I should do and the birds were subsequently liberated in an area of Berkshire called Maidenhead Thicket, a dense mass of bramble. Studying them gave me much pleasure and invaluable information, and I hope they, in return, received a not too unhappy existence at the hands of man.

Cetti's Warbler (*Cettia cetti*)[1****]

General Description
A frequent skulker unless hand-reared, this bird is rarely sighted in the wild. In aviaries it becomes quite fearless when well accustomed to the keeper, but reverts to its skulking habits with the arrival of strangers. It is 14 cm (5½ in) long with a wide tail rounded at the end. The overall colour is a dark rufous-brown; it has light-grey underparts and a reddish tail and a white stripe above the eyes is very conspicuous. The sexes are very much alike but the hen appears a shade duller; the male is whiter on the throat. The song is of outstanding quality, the rich and extremely melodious notes being rendered from a well-hidden site in the undergrowth which it prefers.

Habitat
Generally a resident of marshy areas or their neighbourhood, this migrant hides in dense vegetation and bushy thickets not too far from reed beds. The nest may be found at the base of thick reed clumps on the ground or in dense nearby undergrowth.

Aviary
Although having given these birds, both wild-taken and hand-reared specimens, large aviaries with marshy areas, dense vegetation and overgrown conditions they never nested in my possession (until 1979 while the original manuscript was being prepared for publication). Yet, after wintering another pair and then passing them to a friend, not an experienced bird keeper, who wished to record their song, a pair nested within a fortnight of being placed in an aviary similar in size to that recommended as the minimum suitable enclosure for housing birds. Nevertheless, I would still recommend a larger enclosure if one is to appreciate the general behaviour, including the sudden outpouring of song from one point, then sudden quiet, with the song coming from an entirely different direction within a few moments. A warm shelter or indoor flight for winter use is essential. It need not adjoin the outdoor breeding enclosure but, should it do so, make certain the birds roost indoors during the winter months.

Food

As for most birds frequenting marshy areas or the water's edge, I have always included shrimp meal in their softbill mixture, as well as steamed fish, and roe when possible, to replace the many small crustaceans taken in the wild. This basic food mixture should be rich in animal protein. Fruit can be kept to small amounts but vegetation in the form of dandelion and comfrey should be finely chopped or minced and included in their food daily as it will replace the undigested vegetable matter found in the insects which it would take in the wild. Grated cheese is a favourite addition to the softfood and should be given daily. A very paste-like food can be offered to hand-reared specimens in addition to the drier mixture; it can be based on Farlene (Glaxo) mixed with milk, and a wide variety of vitamin or mineral additives can be stirred into it.

Breeding

The provision of dense vegetation should encourage nesting. Reeds are a simple matter to grow and a bushy hawthorn surrounded by a bed of raspberry canes and reeds would seem ideal. These birds build a base of vegetation and plant stems, and the nest itself is very deep-cupped and constructed from leaves, grasses and rootlets. It is lined with finer grasses, feathers, hair and vegetable down.

Should young result, then offer as wide a variety of insects as possible. When hand-reared the nestlings have readily accepted almost any insects, as have the adult birds kept. With this readiness to partake of most livefoods, nestlings stand an excellent chance of being reared.

Dartford Warbler (*Sylvia undata*)[†]

General Description

This bird has a dark brown mantle, gradually changing to slate-grey on the head. The long fanned tail is white-tipped and is flicked constantly whenever perched. The underparts are pinkish-chestnut to purplish-brown and the throat is flecked with white, with the ventral area white. The length is 12.5 cm (5 in).

Habitat

Wild open country, heath and moorland with low bramble-covered shrubs are favoured. Heather or gorse forms a favourite song perch from which it will dance its courtship serenade, rising almost vertically in the air, then descending while delivering the mating song. One will hear the alarm or scolding note more often than one makes a good sighting. One is lucky to see its shape disappearing as it makes for the cover of a low gorse patch.

Aviary

When serious thought is given to breeding these birds, and to justify honestly their being kept when they are so prone in the wild to decline in numbers, I would suggest a minimum size aviary of 3 or 4 m^2 (9¾ or 13 ft^2) by 2 m (6½ ft) high. The ground should be densely planted with such natural items as gorse, heather and a variety of ferns and bracken, with an area of sandy shingle left in the centre. An unheated but artificially lighted winter shed or indoor flight should be available to them, where their food should be placed to encourage their voluntary entry. Quite often they will roost low in dead bracken or other growth, but the opportunity to retire to warmer roosting sites indoors will not often be ignored. If each corner of the outdoor aviary also has a small roofed section to shield them from heavy downpours of rain, and if they are fed on a suitable diet, these are quite long-lived birds.

Food

As a resident warbler of the British Isles they must be fairly coarse eaters, at least during the colder months. They are simple to feed and will accept almost any small live insect, in addition to the basic insectivorous mixture. Small quantities of soft fruit and minced vegetable can be given with the grated cheese, which they will take in preference to most other items offered. Wax-moth and wood-ant larvae are accepted and will be fed to the young. The parents of one nest, taken in Europe together with their almost fledged nestlings, reared the young almost entirely on these insects and cut-up mealworms in a small indoor flight. Both adult and young took readily to the softfood and cheese provided. I have used small locusts (newly hatched) and a wide variety of insects, and rarely has anything ever been ignored; even minute striped snails have been eaten. Farlene (Glaxo) or similar baby foods, mixed with milk, are also well accepted.

Breeding

My own birds have all been the subject of feeding experiments. In the British Isles this species has become almost extinct on a few occasions and consideration of the food taken in winter time in the wild has always held a strong fascination for me. I have continually pondered the question of whether one can assist in any way by contributing towards their diet during the hard weather when so many losses occur. In the wild the nest is constructed of small particles, leaves, grass, moss, vegetable down, animal wool and hair. Very often the outer walls are decorated with spiders' webs and cocoons. The lining is of fine hair and down. This nest is usually placed very near the ground in overgrown heather, gorse or bramble but is sometimes up to 1 m (3¼ ft) above the ground in the same type of vegetation.

Fan-tailed Warbler (*Cisiculla juncidis*)[†***]

General Description
The plumage of this diminutive bird, whilst very attractive, is not brilliant by any means, but in good feather can be quite striking. The short tail, central feathers, wings and mantle are all blackish-brown and there are chestnut-buff margins to the feathers. The rump and upper tail are rufous, tipped with black; other feathers in the tail are brown but have distinctive broad white tips. The breast is buffish and the flanks have a noticeable rufous wash. The length is 10 cm (4 in).

Habitat
Freshwater areas, marshland, wet areas of bog, agricultural land adjoining open streams, all such sites appeal to this bird. It behaves in rather a skulking manner in low vegetation, for which it has shown an obvious preference on each occasion I have found it.

Aviary
Having watched this species in the wild on a number of occasions, its absolute shyness registered with me and, basically, this is why I chose to hand-rear my specimens rather than to net adults. Nevertheless, my specimens were still unexpectedly shy and, despite invariably feeling rather guilty if a bird remains totally brain-washed and becomes too bonded to me, I was surprised at just how independent they became. This was certainly better suited to my purpose, because accurate and positive behavioural studies could not otherwise have been carried out. Indeed, too many data have already been derived from over-tame specimens and, over the years, many erroneous 'facts' have been passed on in this manner. But even with birds raised as mine were, I would still recommend an aviary 2.5 m (8⅛ ft) long by 1.5 m (4⅞ ft) wide. The enclosure we gave them was well secluded; hops grew on three sides of it and there was a great clump of comfrey, originally planted for finches, which had not been eaten and had reached a height of about 1.5 m (4⅞ ft). There were also decorative grasses up to 1 m (3¼ ft) in height, a bamboo which grew through the enclosure's wire roof, and a flowering

currant (*Ribes*). In the winter, from September onwards, they spent all their time in a bird-room flight a little less than 2 m (6½ ft) long, in a steady temperature of (18° C) (65° F) or just below.

Food

A more-or-less standard warbler mixture was given from the time the birds became self-supporting. With a host of ingredients to select from, such foods can be varied enormously. This warbler is not a great fruit-eater – berries seem to be taken only in late summer or autumn – but even so, grated apple was added to the softfood base daily, as well as the occasional ripe fig or dried figs soaked and mashed; household currants and sultanas were also given. Finely minced steamed ox heart and liver were regular ingredients and, at times, steamed fish and mashed prawns or shrimps; grated cheese was included every day and greenfood was usually chickweed, lettuce, chicory or dandelion finely chopped. The necessary livefood consisted of cut-up mealworms and beetles, wax-moth larvae and adult moths of many kinds. From watching them in the wild, I gather that insects certainly form the greater part of their diet under natural conditions.

Breeding

I ruined a breeding attempt these birds made by allowing a photographer to enter their enclosure. Admittedly I have always liked to achieve first breedings, by no means for the kudos inherent in such achievements but more for anything new that I could learn. I also wished to prove my oft-argued belief that any species of bird will breed under controlled aviary conditions, when kept in a proper manner on a suitable diet. On this occasion, so keen were we to prove this point that it had been agreed to make a short film of the whole proceedings from hatching to fledging; this was to be used for convincing the DoE (Department of the Environment) and any other interested parties that captive breeding was possible. Daily visits did not seem to upset the pair and we were becoming quite confident of them accepting us. We wanted very much to convince those who too often scorned, and even denied first breedings and it seemed an ideal way to combat their attitudes. We could get within just over 1 m (3¼ ft) of this beautiful, hanging, rather pear-shaped nest. Quite deep, it was slung between a clump of variegated grass and the large bamboo; the entrance was at the top and two very pale blue eggs were just visible when we raised ourselves on tip-toe. We had always entered the enclosure when both birds were active away from the nest, and even this time we thought we had seen both of them. However, upon entering the enclosure and approaching the nest, the hen flew from it at our approach. She never again returned to it, even though all future observations were made from outside the aviary. She might have been frightened by the flash when we were taking still shots of her, but we were attempting to get a good shot of the eggs in the deep nest when we entered the aviary. A few days later, when it was obvious that she had deserted the nest and, because of the time of year, would make no further attempts at breeding, I dissected the nest to gather a list of materials used in its composition. It was constructed almost entirely of grass and rootlets and lined with dog hair. The eggs were reddish-spotted on the larger end and the background colour of the eggs now that they were out of the nest was an even lighter blue than it had first seemed when the eggs were sighted in the nest.

Garden Warbler (*Sylvia borin*)[†*****]

General Description
With a length of 14 cm (5½ in), this is a rather undistinguished bird in appearance yet a remarkably fine songster. It is the dullest of our warblers, greyish-brown with darker brown wings, paler throat, breast and underparts fading to white at the vent, with a dark tail and grey-brown legs. The hen is slightly duller and browner. The song is a mellow repertoire of longer duration than most.

Habitat
Rarely if ever seen in gardens except perhaps during migration, it prefers dense bushy vegetation beneath trees as well as wooded areas, heaths and well-brambled moorland.

Aviary
An aviary 2 m (6½ ft) long and high, and 1 m (3¼ ft) wide, will comfortably house a pair of garden warblers. They should have a frost-free enclosure attached and they can be allowed to come and go from the enclosure to the outdoor aviary as they choose, all through the winter months; placing their food and water supplies in the indoor enclosure will almost certainly guarantee that they make full use of it. The aviary should include a densely planted area of hawthorn and raspberry cane, into which long grasses are allowed to intermingle. This should cover at least a third of the enclosure's area and the addition of hops growing at the rear will give extra food supplies natural to their requirements.

Food

At times other than the breeding season, a good insectivorous food supply with a generous amount of fruit – fig, date, banana, apple, pear, grape, all soft fruits, wild or cultured, and wild berries such as elderberries – with a minimum of live insects will keep these birds in excellent condition. Give also a paste-like food as recommended before and include plenty of vegetable matter in their diet. The occasional mealworm cut up on their food mixture will assist in advancing their breeding condition. If a plentiful supply of livefood is available, mealworms need not be increased in quantity but variety can be provided in the form of soft-skinned items such as wax-moth larvae.

Breeding

I would venture to state that the garden warbler is one of the easiest birds of its type to keep in good condition and to breed from. Provide as natural a nest site as possible and offer moss, fine to coarse grasses, rootlets and all animal hairs. Apart from the daily feeding and the occasional offering of insects when in the aviary's vicinity, leave these birds well alone.

Do not be surprised if more than one nest appears, as the male often makes several: whether these are to secure his mate's approval or merely a form of breeding display I do not know. With the arrival of young birds the diet must include a wide variety of insect life; very little fruit will be taken during rearing and indeed the intake of all softfood except the pastes previously mentioned fall to a minimum. As the young leave the nest they will automatically take the paste and other softfoods offered as well as much fruit. With the arrival of migration time it is very surprising how much weight the birds can put on although their fruit intake does seem to grow beyond belief.

Grasshopper Warbler (*Locustella naevia*)[1][*****]

General Description

This bird is 13 cm (5 in) long and of an olive-brown and sombre appearance. It has a well-rounded tail and wings of darker brown, the mantle is streaked darkly and the underparts are lightish-brown. It is the song, or grasshopper-like 'reeling', that reveals its presence more than the rather rare sighting.

Habitat
Very secretive movements characterise this migrant, which is found in open woodland, scrub, grassland with overgrown vegetation, low grassy bushes and even heath and moorland.

Aviary
Provide an enclosure of double the minimum recommended aviary size. Plant two sides densely with thickly growing vegetation and low bushy-type hawthorns through which grasses can grow; to the rear plant hops for seclusion as well as a source of food. Leave a path in between the two beds of vegetation and cover this with gravel as a barrier to weed growth. My own aviary which housed these birds was planted with junipers, spreading horizontally, a buddleia and honeysuckle, and ground covered with rose of Sharon (*Hypericum calycinum*). Winter quarters are essential; extra feeding hours are vital and a little gentle warmth keeps these birds in good overall condition. Such quarters can easily adjoin the summer aviary but, if the birds are allowed out during the day, one should ensure that they roost indoors at night during the colder months when frosts are possible.

Food
Fruit can be kept to a minimum but minced vegetable matter must be added to the basic diet, which can be a high-protein softbill food. Grated cheese should be given liberally and a few cut-up mealworms placed on top of their softfood daily. These birds will eat very young striped snails, smaller than an apple pip; they manage to get the flesh and leave the shells. Ant larvae, the large race preferably, from pine woods or beech plantations, are eagerly taken if provided on the aviary floor. The inclusion of plenty of insects in their diet is essential and culturing one's own supplies is more economical. Elderberries and soft fruits will be taken more in the autumn than at other times, as is usual with migrating warblers.

Breeding
An exceedingly rich animal-protein content in the diet is vital for this bird to achieve breeding condition, which is often indicated by the quivering wings and tail of the male as he proffers a leaf or other small gift to his mate.

If the aviary is well planted so that these birds can travel, rodent-like, in the undergrowth, they will feel at home. Nesting materials to offer are dead leaves, grasses, a little moss, rootlets, bents and a variety of animal hair, fine to coarse. To rear successfully, much livefood is required by the parents. The young are fed solely on insects and their larvae; many of the insects can be cultured for this purpose in adequate quantities. When suddenly increasing the quantity of mealworms with the arrival of newly hatched nestlings, do always ensure they are cut up upon the daily food mixture and not thrown alive and whole to the birds.

Our birds nested, not as I have seen others in the wild, but well into a large clump of grass, hidden completely by overhanging grass blades. Although the nest was cup-shaped, my only way of seeing the contents without disturbing the site unforgivably was to use a dentist's mirror. The nest itself must have been at least 40 cm (16 in) above ground level as the clump of coarse grass had been planted in a deep plastic bucket. The hen produced six glossy white eggs with purplish-brown markings almost covering their surface; of these only three hatched. I firmly believe that, had an old friend from Sweden not visited us at that time, bringing a large supply of wood-ant eggs, I would have had great difficulty in feeding the nestlings. My own wax-moth larvae were far too small, although, in the later days prior to fledging, many of these were taken for the nestlings. The young fledged at 13 days. When dissected, the nest was found to be composed only of dry grass of a very fine nature, thin stalks of dead vegetation and dried chickweed, with both dog and cow hair for the lining. Even though a true cup shape, the entrance through the surrounding grass was very tunnel-like and extremely small.

Icterine Warbler (*Hippolais icterina*)[1*****]

General Description
Very like a large wood warbler, the icterine warbler is stockily built. It has yellow underparts and olive-green on the mantle and wings, although the wings and the tail are a shade browner. The legs are bluish-grey and the inside of the bill is red and most noticeable when taking food from the hand. It has a habit of raising the crown feathers when curious. It is a very lively bird, flitting among tree foliage in its search for insects. The song is rather a discordant jumble of notes, sometimes liquid-sounding and sweet to the ear and at other times quite harsh. The length is 13.5 cm (5¼ in).

Habitat
Never very far from human habitation, this migrant is found along river banks among tall trees, on the edge of woodland, in orchards, tall hedgerows and even gardens where fruit trees and bushes abound.

Aviary
An enclosure of 3–4 m (9¾–13 ft) long by 1 or 2 m (3¼ or 6½ ft) wide and 2 m (6½ ft) high would suit this energetic warbler. Plant it with hops along one long side and grow a thick hawthorn hedge at one end; into this let plenty of raspberry canes grow. Provide a roofed portion over the top of the hedge, just large enough to shield a nest from the main force of the rain but allowing adequate water to reach the growing vegetation, and the birds will be encouraged to use the hedge for nesting. I have always found this bird to like a little gentle heat during winter months in addition to extra feeding time, so the shed-shelter or indoor flight, perhaps attached to the outdoor aviary, should be equipped with some type of heating apparatus.

Food
A high-protein insectivorous mixture, with plenty of both fruit and vegetable in the diet should be supplied. These birds seemed so very fond of eating butterfly eggs that I mixed fish roe in with their food, this being very similar and an extremely high-protein content food. Farlene (Glaxo) will also be taken, as will almost any small insect; wax-moth larvae, which quite possibly resemble a natural food of theirs, are a firm favourite at all times. When mealworms are fed, ensure they are cut up and never given alive. Elderberry and a number of other wild and garden-grown soft fruit are eaten and, in fact, I have frozen supplies during the summer and autumn to be given later when they were out of season.

Breeding
Normally nesting in the fork of bush or tree, the recommended site described under the aviary heading should prove suitable. It is an attractive nest composed of grasses and rootlets all tightly interwoven with a final lining of fine grass and animal hair. Rearing of any young will depend entirely upon the ability of the bird keeper to provide ample small insect life.

Melodious Warbler (*Hippolais polyglotta*) **

General Description
Rather like the icterine warbler in appearance, but perhaps just a shade smaller, this bird is without doubt a much better songster to my ears. The mantle and wings are predominantly olive green and there is far more yellow on the underparts.

Habitat
It is found on river banks, by open woodland streams, in orchards and even in hedgerows in close proximity to water.

Aviary
During the spring, summer and autumn of the year following that in which we hand-reared them, our birds lived in a rather densely overgrown enclosure, some 2 m (6½ ft) long by 1 m (3¼ ft) wide. At one end grew a thick privet that had become entwined with hop growth. Knowing of their fondness for water, we provided an old earthenware sink to act as a small pond for their entertainment. Elsewhere we had encouraged a strong growth of nettles. These, and a comfrey plant a number of years old and subsequently very thick, covered almost half the aviary, making entry neither easy nor pleasant.

Food
Avi-Vite formed the basis of these birds' diet. To this was daily added a wide variety of foods: always grated cheese, minced steamed ox heart or liver, fish – made acceptable to them by steaming, mashed boiled potato in small quantities, hard-boiled egg, finely chopped prawns or shrimps, chopped lettuce or other similar greenfood – even chickweed is acceptable. For live food we offered mealworms cut into pieces, wax-moth larvae, moths, spiders etc.

Breeding
No attempt to nest was ever observed. The female always seemed to be present when the aviary was checked and the male appeared to spend most days serenading her with no response for his labours. Finally, in September, when catching them for transfer to their winter quarters – an indoor bird-room flight, two deep nests were discovered woven into both hops and privet, but only just below the surface of the vegetation, beneath rather large hop leaves. In fact so tightly were these woven into position that they were badly distorted in recovering them for dissection. Neither nest held eggs but, in taking them apart, we found egg-shell fragments in one; although the nests had been on the outer branches of the privet they had been completely hidden from sight. These nests were constructed of chickweed, grass, roots, dog hair and fur, most likely from rabbits since we had offered a variety of materials during the season. It was obvious that mice had entered the aviary at some period during the summer, for, beneath the privet hedge, where the ground was hard and dry, were a few mouseholes. The presence of these pests may well have accounted for the lack of eggs, or even for the reluctance of the birds to

show any further interest; but this is theory only, with no proof to back it up.

Orphean Warbler (*Sylvia hortensis*)

General Description
A rather large bird, fuller in the body than the blackcap, the male has a jet-black crown but the black extends down to well below the eye, giving it black cheeks. The throat is a conspicuous white, as are the outer tail feathers. The mantle is greyish. Its length is 15 cm (6 in). It has a rather repetitive song. The female of this species is similar in appearance to the male but without the density of black about the head and face, being more dull in this area; the breast too is slightly paler on the whole.

Habitat
Open woodland, orchard, park, scrub, olive groves and even gardens are visited by this migrant. It spends much of its time flitting through the trees taking insects from foliage and while on the wing.

Aviary
This member of the warbler family requires a maximum of flying space. Aim for a length of 4 m (13 ft) with a minimum of 1 m (3¼ ft) width and 2 m (6½ ft) height. In the wild this bird will nest up to well over this height in a tree or thick bush. Cultivate a thick hawthorn bush in the rear of the enclosure with a dense mass of raspberry canes growing up through it. I have kept these birds only for experimental feeding purposes, never with the intention of trying to breed from them. However, having seen them nesting in the wild, I would think this type of site should be acceptable. Separate warm winter quarters are necessary.

Food
I have kept freshly taken adult specimens and also hand-reared this species and they seem extremely simple to cater for. Offer the usual insectivorous mixture and grated cheese, with plenty of fruit and vegetable matter, the latter being minced in the softfood. This is another bird which readily takes Farlene (Glaxo) mixed with milk. The fruit can consist of ripe apple, pear or almost any soft fruit. In late summer, elderberries and blackberries will be appreciated. Cut-up mealworms in limited quantities can be given along with most small soft insects; flies, moths and larvae will be accepted but my birds were never very keen on locusts.

Breeding

A tall, thickly growing hedge in the aviary will provide a nest site similar to that chosen in the wild and, when looking for these birds, I have found nests in branches between 2 and 3 m (6½ and 9¾ ft) from the ground. These nests are made from grasses, weed stems, fine twigs and rootlets, decorated externally with spiders' webs and cocoons and lined with hair.

When watching a pair feed their young in the wild I saw a vast number of unidentifiable small caterpillars being given. When keeping them I have tried them with both wax-moth and wood-ant larvae; these were taken in very large quantities and I imagine they would be able to rear young on these as other warblers in my possession have done.

Reed Warbler (*Acrocephalus scirpaceus*)[†*****]

General Description

This bird has a rather erect stance with a slightly raised crown. It has a rufous-brown back, a tail of darkish-brown and a crown of brown with a distinguishable rufous wash; the rump too bears this same shade. The throat, breast and belly to vent are much paler – a soft beige-biscuit colour. It is, therefore, not by any means a colourful species but the male is one of the noisiest of birds during courtship, with a continuous and rather monotonous song. Sometimes it barely ceases throughout the day and is broken occasionally only by a few 'chuckling' notes. The song is often the only means by which the sexes can be told apart. It is 12.5 cm (5 in) in length.

Habitat

This migrant is rarely found very far from reed beds, or smaller ditches with sufficient reed growth to provide a nesting site. At times it is forced to congregate in small areas due to a shortage of nest sites.

Aviary

On the only occasions that I have known these birds to nest and rear successfully, their aviary included a pond and small reed bed; plastic buckets, an old-type square earthenware sink or even a garden fish pond incorporated in the enclosure will allow plenty of reed growth to meet their needs. The aviary itself can be as small as the minimum standard breeding unit. The inclusion of willow, planted at each end, gives a near natural setting and ensures a maximum of flight. One of the vantage points provided by this vegetation will become a favourite singing site where the male will carry out his courtship song, fluffing out his feathers and fanning his tail. During the winter months these birds will require a shed-type shelter to which they can retire when too cold outside. Extra feeding time, with the provision of artificial light, is a more vital factor for good general health than extra warmth. Instead of this type of shelter one can remove them to indoor flights. They may normally be housed together all the year round but, to house two pairs of these birds, an aviary would have to be at least 7 m (23 ft) long with a pond in the centre and a feeding area at each side. There should be an osier bed at each end of the enclosure. A little outdoor roofed shelter is needed to cover a few of the rear willow branches.

Food
A good-quality insectivorous mixture must form the basis of the diet and to this can be added a few cut-up mealworms and grated cheese (Cheddar-type consistency). Free access to the pond will ensure that a number of natural insect foods are taken.

Breeding
The nest of this bird is a wonderful piece of engineering. It is slung between growing reeds, often only just above the water surface, and is cup-shaped and made from grasses, moss and reed flowers, lined with feathers, hair and wool or vegetable down. Quite frequently the hen will slide off the nest and the male will take her place without any indication other than the song ceasing. When rearing the parents will need much livefood.

Great Reed Warbler (*Acrocephalus arundinaceus*)**

General Description
Resembling a giant reed warbler, this bird measures 19 cm (7½ in) in length. It has a conspicuous eye stripe and a far stronger bill and, although the song is similar, it is far louder and seems almost as loud as that of the thrush when one is standing near the aviary. As the song includes many coarse grating and croaking notes, one can be thankful that it does not continue to the same extent as that of the smaller birds. Nevertheless it is a fascinating bird to keep, becoming delightfully tame in a very short while, with such confidence that it will follow one about to obtain the luxury of a mealworm from the hand. The sexing of these birds can be difficult, but the noisy specimen is invariably the male.

Habitat
This migrant frequents freshwater margins, predominantly reed-bed areas throughout the European mainland.

Aviary
During the out-of-breeding season months I have housed three pairs together with only the slightest of bickering. As with its close relative it requires extra feeding time and a draughtproof indoor flight or shelter. The aviary should be similar in all respects to that of the smaller bird.

Food
I have seen adult specimens of this bird take a 15 cm (4 in) locust and hammer it until limp and fit to swallow. They will readily kill and dismember mealworm beetles which have been discarded after their egg laying usefulness is finished and, from this, one can appreciate that they have a taste for livefood of many kinds. A good insectivorous mixture is taken freely, along with grated cheese. Many angling friends used to pass over their catches to me so that we could steam and mince the fish and include this in our softbill foods.

Breeding
I can really add little to that already advised for the more common reed warbler (page 158). This larger member of the warbler family can be treated in an identical manner at all times.

Sardinian Warbler (*Sylvia melanocephala*)[†*]

General Description
The behaviour is very like that of the lesser whitethroat, and the underparts are also similar, but here the likeness ends. It has a conspicuous eye and eye ring of red; the back, wings and tail are grey and the dark, almost black, tail is graduated at the sides with white outer feathers. The head of the male is black, and this extending below the eye to the cheeks. The female is less dark with no black cap and she too closely resembles the female lesser whitethroat. The male has a dancing display flight. The song, otherwise very whitethroat-like, is carried on over longer periods and is perhaps a little more liquid in its rendering.

Habitat
A woodland bird, this migrant frequently skulks in densely growing vegetation, bramble or bush, but tends towards more open park-type

country. It is often seen for a few moments in country lanes in Mediterranean countries before diving for cover with its scolding, rather wren-like chattering.

Aviary
If half of the aviary is densely covered with vegetation such as raspberry canes, honeysuckle and bracken, or dense evergreen shrubs, the standard minimum size will suffice; leave the rest of the floor area bare or with low-growing heathers, rockery plants and such-like. Winter quarters will be necessary, frost-proof and possibly warmed, but remember that more important is the extra feeding time made available by the use of artificial lighting. This can be operated simply by using a time-switch so that the birds' light comes on at 7 a.m. until daylight then again later in the day until 8 or 9 p.m. Where the winter housing does not adjoin the aviary, the aviary will need to have a small roofed portion providing shelter from heavy rainfall.

Food
A good softbill mixture should be the base to which one adds grated cheese, fruit of all kinds and some vegetable matter. I have witnessed these birds taking small buds from nettles and even tearing a leaf of chickweed to pieces in their feeding habits. Fruit is essential, so always keep ripe apple available; wild items such as elderberry, blackberries etc. will be readily accepted. In fact, when fresh elderberries are made available, very little else is taken.

Breeding
These birds normally nest very low, almost at ground level, in the wild but where the natural vegetation of the area does not allow this, I have seen nests as high as 2 m (6½ ft) and more. They prefer a dense mass of growth; honeysuckle was selected by my birds on two occasions, despite a wide variety of other vegetation being present. Materials needed are bents, coarse and fine grasses, rootlets, plant down and hair. Externally the nest looks untidy but the inner cup is beautifully finished; the outside is often decorated with spiders' webs, some even being used on the internal structure.

I cannot over-emphasise the value of the larvae of both wood ant and wax-moth when young are newly hatched; the first are obtainable in the wild in great numbers and the second very easily cultured for this specific purpose. Even newly hatched locusts will be taken but whatever is offered as the basic livefood should be persevered with throughout the rearing process; continuity of rearing foods is very important.

Sedge Warbler (*Acrocephalus schoenobaenus*)[1,****]

General Description
The song alone makes this bird a welcome addition to many a collection. A healthy contented male with a mate somewhere in the background will sing almost from dawn to dusk from his choice of perch. Although it can be described as brown, this does not do justice to the feather pattern which is of a warm brown, darkly streaked on the crown and back. There is a broad whitish eye stripe, the rump is rufous and the tail is dark brown. The breast and underparts are a creamy hue, darkening on the flanks and it is grey legged. It measures about 13 cm (5 in) long.

Habitat
Lakes, ponds, rivers and streams attract this migrant, the vegetation on the bank or nearby low hedgerow providing its nesting site. It is seldom visible for more than a moment or two as it flies from one song position to another and disappears in the reeds. Its flight is low and of rather a jerky nature.

Aviary
Use the standard minimum-sized aviary. Include a pond to grow reeds and provide a clear area of water. If possible obtain two or three large clumps of sedge and plant this around the pond in such a manner that it keeps moist; the over-flow from rainfall will suffice to keep this grass growing and it will usually form into at least one nest site. A couple of teazle plants with long grasses growing between them will give some attractive song perches and nest sites, adding eye-catching vegetation to the enclosure with its blooms. A warm shed-like shelter or indoor flight is necessary, together with a few hours of artificial lighting during the darker days of winter to enable later feeding. The maximum hours of darkness I allow my stock, even in an unheated indoor flight is 10, but for some this may be too long – with the aid of a nursery-type light I have witnessed small insectivorous species eating at midnight.

Food
For most of the aquatic warblers I often include a few minced small crustaceans, such as prawns and shrimps, in their diets to replace some of the small crustaceans that they obviously take in the wild. The basic diet is still an insectivorous food of high-protein value. To this is added the usual grated cheese. I give these birds very little fruit, but increase the vegetable matter, such as minced dandelion, comfrey, carrot and brassica. Sedge warblers will take almost any small insects, from ants'

larvae to locusts. This is another bird that I partially feed on Farlene (Glaxo). For many of the smaller warblers I include a few drops of Abidec (Parke, Davis) or similar multivitamin liquids.

Breeding
One can offer nesting sites as mentioned in the description of the aviary. Feed the stock well and hope that they will nest and rear. Offer materials such as weed stalks, moss, grasses, hair, vegetable down (willow) and feathers. Any young will require much livefood so plan well beforehand, perhaps by culturing various insect food items. I know a bird keeper who grows a variety of vegetables for his own use in his aviaries, firmly convinced that, since they encourage aphids, the birds will keep his plants insect-free and benefit at the same time.

Subalpine Warbler (*Sylvia cantillans*)****

General Description
Resembling rather a paler version of the Dartford warbler, the male bird has a white moustachial stripe. His shorter tail has white outer feathers and his overall length is 12 cm (4¾ in). The upper parts and mantle are pale grey whilst the wings and tail are considerably darker and the breast is rufous-pink. The female is far duller in appearance, predominantly of greys and light browns.

Habitat
Thick vegetation and scrub is the favourite habitat of this migrant warbler, as it is of the whitethroat. Its movements too are very whitethroat-like, as is the dancing song-flight display.

Aviary
The minimum-sized standard enclosure is large enough for a pair of these birds. Plant life should be as for whitethroats (page 166) or Sardinian warblers (page 160). Winter housing needs to be warm and frost-free, and extra feeding hours are vitally important.

Food
A rich protein food should have grated cheese, fruit and vegetable such as dandelion grated finely and added. The whole should be rather dry, not too moist. These birds will, however, accept a supplement dish, consisting of Farlene (Glaxo) baby food moistened with milk to a wet paste.

Breeding

This bird will often nest in creepers as well as low dense shrub-like vegetation. Materials to provide are coarse and fine grasses, animal hair and plant down. The male will frequently be seen displaying to the hen whilst holding plant down in his bill.

Any young will need much small livefood, as does the Sardinian warbler. They readily accept almost any soft-skinned insect for rearing purposes, certainly showing a preference for moth larvae, wood-ant larvae and newly hatched locusts.

Willow Warbler (*Phylloscopus trochilus*)[1]*****

General Description

This bird is greenish above and more yellow beneath; the legs are light brown to flesh-coloured. The length is 11 cm (4¼ in). It can easily be confused with a number of other warblers, the most common being the chiffchaff, although the legs of the latter bird are always dark. The eye stripe in both birds is yellowish. Alike also are the call and alarm notes but the song is totally dissimilar. Young willow warblers are very yellow and in the wild can be easily mistaken for wood warblers.

Habitat

Scattered trees, woodland dells, quiet secluded gardens with trees, bushes and fruit canes, all provide this migrant with the small insects which form its diet. I would hazard a guess that, of the warbler family, it is the most common summer visitor to the British Isles.

Aviary

Use the standard minimum-sized enclosure and house the pair on their own but together all the year round. Willow warblers will need to be wintered indoors in a flight or have a shelter attached to their aviary; they will frequently venture out during colder months but, with their food placed in the shelter and extra feeding time given, they will rarely attempt roosting outside. If the shed is illuminated they will enter at dusk and remain in its draught-proof safety overnight. Give thick bramble-like bushes.

Food

An insectivorous mixture, grated cheese, ripe apple, pear, peach and fig or avocado pear should be offered; some fruit should always be present. A little Farlene (Glaxo) may be given daily mixed with milk to which can be added vegetable matter. Larvae of the wax-moth, so easily cultured and given in addition to the above foods, will keep these birds healthy for a number of years; mine rarely saw a mealworm and seemed not to suffer from its exclusion from their diet.

Breeding
The aviary in which I bred these birds was densely overgrown. The bulk of the aphids they took whilst rearing came from a hop plant that covered the rear of their enclosure but, since it contained so many plants and flowers there were probably many types of insects I was unaware of. Provide grasses, dead leaves, mosses, fine rootlets and twigs with hair and feathers for nest lining; my own birds even used skin from honeysuckle and what appeared to be dried leaves from garden iris. On the occasion when these birds reared successfully in my aviaries I was able to provide wood-ant larvae together with wax-moth larvae; these, together with the vast amount of aphids present, gave the parents most of the insect life needed.

Wood Warbler (*Phylloscopus sibilatrix*)***

General Description
This bird has a more distinct yellow eye stripe than either the willow warbler or chiffchaff and is also larger, measuring nearly 13 cm (5 in). In body bulk, too, one can notice the difference. It can sometimes be confused with a young willow warbler due to the similar amount of yellow exhibited: the legs are yellow and the breast, also of this colour, is very noticeable. The mantle is yellow-brown with an olive-green wash to it and the underparts are white. The wings are longish, making the tail appear shorter than it is.

Habitat
Beech woods and open forest areas are the favourite haunts of this migrant and, although in some parts of Europe quite common, it is often not seen and one is only aware of the bird's presence by the song or catching sight of its frequent aerial excursions after insects. One might see the attractive mating display, where the male will behave like a large dragonfly, vibrating the wings in a blur of speed and ending in a spiral descent in his wooing of the female. These birds nest in woodlands amid overgrown brambles and the dead branches from the overhead trees or in thick dead bracken newly surrounded by the current year's growth. Sometimes they make use of a depression in the ground. The nest itself

is made of dead bracken fronds, leaves, grass and bents and lined with hair.

Aviary
A little above-average height should be the aim for an enclosure to house these birds. Without this one cannot really appreciate the full beauty of their movements during the breeding season. The absence of extra height may also limit the display which is needed to trigger off the hen's response. Just over 2 m (6½ ft) in height, or an apex-type roof of this maximum, should suffice. A length of 2 m (6½ ft) and a width of certainly no less than 1 m (3¼ ft) would be required, the width being increased to 2 m (6½ ft) if possible.

Plant the aviary thickly at one end, or at the rear, with both brambles and bracken; if brambles are considered too unpleasant to negotiate when entering the enclosure, substitute raspberry canes as thickly as can be managed; these will hold their own with the bracken and provide near-ideal conditions. When planting the aviary in this manner, leave half of the floor area as peat and shingle covered with heathers to restrict weed growth. A frost-proof winter enclosure will be needed and a little gentle warmth may be required if the birds seem to feel the cold. The extra feeding time provided by artificial light is essential for general good health and longevity.

Food
A good-quality softbill food containing a high percentage of animal protein should form the basic diet. One should add plenty of grated cheese, sweet apple or pear and finely minced greenfood; mine received daily rations of both wax-moth larvae and cut-up mealworms on their food. The former were given to them alive. During the autumn months there is an abundance of wild berries to be taken; such fruit as elderberries, raspberries, blackberries, and even the soft fruit of rowan, can be given; I have seen these birds eat the flesh of rowan that has been lying beneath the trees after finches and the like had removed the pips and discarded the soft flesh.

Breeding
As mentioned in the section on habitat a natural site would be dense bracken and undergrowth; almost any similar type of site will be accepted by these birds once in peak breeding condition. Much of their nesting material will be collected from the surrounding site: dog hair or other suitable lining material will have to be added.

Since much of their food is caught on the wing when at liberty, I used to liberate a daily supply of wax-moths into their enclosure. Remember that the fly resulting from the fishing maggot is little else than roughage unless it has been specially fed. If feeding these in any bulk at all firstly ensure that they have access to honey-water to which has been added a vitamin supplement. When rearing is in progress the birds will take small locusts, grasshoppers, moth larvae, cut-up mealworms, maggots treated with Vionate (Squibbs) or similar additive, earwigs and wasp or bee pupae. The young we hatched were reared for the first week upon wax-moth larvae and wood-ant larvae, and as the adult pair had been reared by hand on a Farlene-based food mixed with milk, they did at this time offer their young a certain amount of this valuable food.

Lesser Whitethroat (*Sylvia curruca*)***

General Description
This bird has light, silvery-grey upper parts, darker ear coverts and cheeks, and a snow-white throat but pale pearl-grey on chest and belly fading to white at vent. The outer tail feathers are white. The length is 13.5 cm (5¼ in).

Habitat
Open woodland with plenty of dense undergrowth, brambles and nettles is particularly attractive to this migrant, along with orchard areas which have been neglected over the years. Nests are about 1 m (3¼ ft) from the ground in dense bush or bramble.

Aviary
These birds can be housed together all the year. The standard recommended minimum-sized aviary is ideal for one pair of these birds. Ensure that a shelter is attached or available for winter use; a little gentle warmth will be required at that time as well as additional feeding time. The aviary should be planted thickly at one end with a densely growing hawthorn; beneath this, and growing in and around it, plant plenty of raspberry canes; clusters of dead bracken can be attached in the corners at various heights. There should be a small roofed section at each end. Supply a small shallow pond as they are fond of bathing often. Keep the ground growth low, apart from in the rear nesting section.

Food
As for the willow warbler (page 163), but include more fruit in the diet.

Breeding
Materials such as fine twigs, grasses, rootlets, dead leaves, spiders' webs, vegetable down, animal hair and sometimes small pieces of fluff rather like tiny pieces of cotton-wool should all be offered.

When the eggs hatch the young are small and their requirements are correspondingly so. It is no use at this time to offer maggots; later, at the 8-day stage, they will be taken but now they need aphids, wood-ant larvae, moth larvae and a supply of mealworms, increased daily, to cope with the urgent need of the nestlings. Large transparent plastic sacks can be very useful at this time: whereas in the old days we used an upturned umbrella to collect insects from tree branches, one can now put small branches of beech or birch almost fully into these sacks and shake the branch well, collecting over a short period a host of minute insects to release in the breeding pen or freeze for later use.

Whitethroat (*Sylvia communis*)****

General Description
This bird has a rufous-brown back and wings and a greyish-brown head. The underparts are light buff and delicately tinged with pink, the throat is very white as with the lesser whitethroat and the tail's outer feathers are white. The length of this bird is 14 cm (5½ in). The sexes are very similar; if anything the female is browner on the head and a little more dull in the body, with no pink tinge.

Habitat
Dense nettles, brambles and other undergrowth attract the migrant whitethroat. It is not so skulking of habit as the lesser whitethroat. The nest is usually placed low in thick hedge, bramble or plant.

Aviary
As for the lesser whitethroat (page 166).

Food
As for the willow warbler (page 163) but include more fruit.

Breeding
As for the lesser whitethroat (above).

9

Flycatchers, Tits, Bearded Reedling and Wren

OF THE OTHER SMALL INSECTIVOROUS species grouped here the flycatchers are altogether more delicate and difficult to maintain in the best of condition over many years. Very serious consideration must be given as to whether one can devote the time and energy which will be required to look after these birds properly. The spotted flycatcher is really no more troublesome to care for than most insectivorous species, but the other species mentioned are very much more difficult to keep.

Whereas livefood is almost vital for all these birds in rearing young when in the wild, one can hand-rear excellent specimens without it, replacing with other easily procurable animal protein the small insects given by parent birds in the wild. Some hand-reared birds, if kept on a diet similar to that on which they were raised, will in turn offer such food to their own offspring, but unfortunately this is not a frequent occurrence. We tend in general to offer a softfood and then do everything possible to distract the birds from it by providing much livefood elsewhere in their aviary.

It is far better to have one food dish only and into this should go the cut-up mealworms, wood-ant pupae and wax-moth larvae, all mixed well into the basic food. The provision of growing hops or other vegetation in the outdoor breeding enclosure will offer a wide alternative variety of insect life. A wide dish with a covering of food so shallow that there is little need for the birds to throw it over the sides will be found very much better than deep food vessels, in which the bottom layers of food will remain untouched.

We must offer a rich diet to flycatchers, equivalent to that of their wild-insect food intake. We should pause and dwell for a moment on just how wide the variety of insect life is and what the undigested matter in them is likely to be: pollen, liquids, unlaid eggs or even smaller insects.

Although we may not be able to equal this variety we can ensure quite simply that the basic softfood is fresh, nutritious and wholesome and, a most important point, palatable. They will need a small amount of fruit and fresh vegetable matter but for these birds it is more important to include steamed fish, fish roe, ox heart and liver, shrimps and prawns, all minced. If very small fish, such as minnows, are available they can be killed and passed through a mincer before being stirred into the basic mixture. A man-made softfood can be excellent as a basic food, but fresh items cannot be stored for long and it is in the provision of these that the main care is needed for these birds. The softfood is a basic requirement and, although it should be adequate for a bird to live on temporarily, it was never intended to meet each and every bodily need.

It is not essential for every bird keeper also to be a budding nutritional chemist or even to study this sphere in any depth. As long as he knows which foods supply suitable protein, vitamins, mineral salts, fuel foods for slow and quick body absorption and roughage, he should be able to supply all birds' needs. Fresh water is one of the most vital requirements and should be given in addition to the many liquids present as moisture content in the solids. The recommended items are not intended to be fed in large quantities: small amounts over a long period are far more beneficial. The only item that I always use regularly is grated Cheddar cheese. If one or two other items are missed from the diet for a day it then matters little.

A healthy bird is one that clears the food dish daily, taking the various foods as they come and not forever searching for an item that is in short supply or not present. The extra livefood taken from the wild by a bird such as this will not deter it from taking the basic diet.

The other birds in this group *need* not be fed in quite so specialised a manner, although personally I like to standardise when I can and treat most of my small insectivorous birds in this way; the larger ones have a slightly more coarse diet.

When young are being reared and the problem arises of livefood supplies failing to last until the keeper can replenish them, one or more 15 cm (6 in) deep trays, up to 1 m^2 (3¼ ft^2), will provide the answer by keeping some insects available for many hours. Line the inner sides of the trays with aluminum kitchen foil, which can quite easily be stuck to the wood. Then add bran to a depth of 5 cm (2 in), followed by the insects – mealworms, buffalo worms, wax-moth larvae, crickets etc., mixing these gently into the bran. The vast majority will now be hidden but, every now and again, one of the creatures will surface or the parent birds will turn over the contents by natural hunting methods. The foil will prevent escapees. Using this simple system of regulating access to food not only obviates the too-soon exhaustion of the daily supply but also provides exercise and encourages the male to hunt in a natural manner, thus harnessing his energies into a useful pursuit. Food can be found whenever the birds need it and there is no danger of the supply of insects running out perhaps for hours. Do make every effort to house these trays in a dry area, clear of any rainfall; otherwise hunting may become difficult and escaped insects can hide away unseen. Many insectivores eat considerable amounts of grain if it is ground to a coarse flour; pigeon conditioning mixture, together with legumes and peanuts, can be passed through a coffee grinder. When placed in a separate food dish, it will be seen to receive constant visits and may only need replenishing every few days; no longer than this should be allowed to elapse as a number of foods, once ground, can develop a fungal growth, which is of a toxic nature.

Red-breasted Flycatcher (*Ficedula parva*)

General Description
A mere 11.5 cm (4½ in) long, rather like a slim robin, this bird has an ever-flicking tail which exhibits to advantage the white patches beneath which are just visible from the side view. It is mainly brown above and white below. The male only has the red breast, his mate being more of a pinkish-biscuit shade but pale in this area. The song is a chattering, wren-like warbling.

Habitat
Woodlands, broad-leaved forestation and parkland are the haunts of this migrant. It is rarely seen except when it leaves the cover of foliage to take insects on the wing; then the white under-tail patches are conspicuous.

Aviary
I would never consider an enclosure smaller than 2 m^3 (6½ ft^3) as suitable for this species. My own birds were hand-reared and consequently tame but, even so, nervousness showed itself when strangers came to view these birds. Apart from this, the beauty of their flight could not be seen or the necessary exercise taken in a smaller aviary. Plant beech along two sides of the aviary and keep this stunted by cutting the tap-root for the first few years. Winter quarters of a warm nature should be either attached or otherwise at their disposal for the cold months, but I have allowed them free use of outdoor aviaries, making certain that

they roosted indoors at night. These birds often retire to the seclusion of an attached shed with the first signs of strangers approaching, even in the summer.

Food
With some flycatchers, including this one, I like to add steamed ox heart and liver to a good basic softfood. Still provide the grated cheese, miss out the fruit except when fresh elderberries are available, add some finely minced or chopped vegetables and, on top of the food, put four chopped-up mealworms per bird per day. Wax-moth larvae should be given in small quantities but it is essential that one does not over-do the livefood intake to the point where the birds do not take the insectivorous mixture. I have seen some inexperienced bird keepers hatch out 4 litres or so (about ½ gallon) of maggots into flies and proudly exclaim how their flycatchers are living on them; it has often been too late to rectify the damage done. Although they are not normally fish-eaters, I have included steamed fish roe in their food with excellent results; this roe of course resembles the eggs of a number of insects and as such look like a natural food. Very often it is the sudden increase in livefood availability that upsets rearing birds; this sudden change of diet is not necessary with hatching of nestlings. For the first day of their lives they require little insect life. Having reared from the egg by hand, I am well aware that the first few hours are essentially a drying-off period when warmth is the vital ingredient to success; the food intake at this time is of the barest minimum.

Breeding
These birds usually nest in the hole of a tree or a brick wall but their nests have been found against the bole of a tree with small twigs or branches holding the structure in position. Offer a variety of nest boxes, facing in differing directions. They can be of two distinct types, either half-open-fronted or with just an entrance hole. If using the latter make sure that the hole is a very large one; strangely these birds seem to avoid small entrances. The nesting materials are quite varied: moss, lichen, fine twigs, dead leaves, plant down, animal hair and spiders' webs will be used.

The larvae of the wood ant, wax-moth, or any other small moth acceptable to them, day-old locusts and wasp and bee grubs, should be offered. Supply as wide a variety as possible and see which are used most for feeding their offspring. Whereas the pied and white-collared flycatchers can often be seen taking small beetles, the red-breasted variety seldom if ever seems to eat them; it certainly has a more delicate-looking bill and, like the spotted flycatcher, spends much time picking-off flying insects, flying from a favourite perch and returning with their prey. Unless one manages to obtain hand-reared specimens, I doubt if these birds will take to the baby-food paste previously recommended as a nutritious supplement to their diet; those which have been reared by hand usually retain a taste for foods which were introduced during their rearing.

Spotted Flycatcher (*Muscicapa striata*) ***

General Description
It has a grey-brown mantle and the wings and tail are darkly striated. The breast is paler but dark streaks cover the upper chest fading to white on the underparts. The bill is greyish-brown and the legs are black. The sexes are alike except for a slight difference in the throat marking: the male has a larger white area here.

Habitat
Sparse woodland areas, orchards, parks, gardens, a small groups of trees or a country lane edged by trees will attract a pair of these migrants for their short breeding season. They will be seen to fly out from the foliage, catch some flying insect and return to their selected perch. They spend hours feeding in this way, but rarely feed at ground level.

Aviary
The recommended minimum-sized enclosure will house a pair of these birds. There is no great need for a lot of vegetation, except perhaps for that which encourages insect life. I have known them to nest in dense-growing hops covering the side of their aviary, but equally they are capable of choosing a shelf or open-fronted nest box as a site. Their winter quarters must be frost-free and can be either attached to the outdoor breeding pen or a separate building to which they must be moved; the former is always to be preferred. A little gentle warmth may be necessary if they show signs of feeling the cold. If the shelter for winter use is not attached, some form of roofing will be needed during inclement weather.

Food
A good insectivorous mixture must form the base, with various additions such as grated cheese, a little finely minced or grated vegetable, a few cut-up mealworms daily and even freshly killed beetles. Where flies are

added to the diet, keep the numbers down or make sure that a dish or saucer of honey-water is available for the flies to drink from; otherwise they are mere roughage, with little in the way of any nutritional content.

Breeding
Usually the nest is to be found at a height of 2 or 3 m (6½ or 9¾ ft) so a high nesting site in the aviary will normally be preferred. The materials required will be moss, grass, rootlets, fine twigs, fibrous plant skins, spiders' webs, plant down, feathers and hair.

These birds are most obliging in that they accept most insects when rearing. However, if one is supplying maggots then a dusting of Vionate (Squibbs) and extra calcium lactate will be required. Still provide cut-up mealworms on the food, and the larvae of ants and moths if possible.

Pied Flycatcher (*Ficedula hypoleuca*)[1***]

General Description
The male is strikingly black and white during the breeding season but the female is at all times far duller in plumage, being merely a grey version of him although she retains the white wing bar. During the winter months a male will very much resemble the female but he still exhibits a slight whitish forehead and, upon really close inspection, a little more white in the wing bar. This pied species of flycatcher can be hardly distinguishable from its white-collared relative after the autumn moult, though at close quarters one can just make out the dull remains of the collar of the latter bird and what is left of the white rump, both these areas being faintly visible as grey-white when compared with the true grey of the pied flycatcher; this, of course, happens only during the 'eclipse' plumage. During the breeding season the male is black on the crown, cheeks, mantle, back and tail and has a white wing bar, forehead, breast and belly. The female exhibits grey where the male is black, and her wings are slightly darker than her mantle but retain a white wing bar. The length of this bird is 13 cm (5 in).

Habitat
Forests, woods, orchards, anywhere with an abundance of trees will attract this migrant. Unlike the red-breasted or spotted flycatchers, this bird often feeds upon the ground; fallen caterpillars and small beetles are more often than not the food taken in this manner. The nest is usually made in the hole of a tree or even in a nest box provided by man. Many

birds build their nests in holes in walls, the materials used being moss, grasses, dead leaves and hair.

Aviary
Make an effort to provide a large aviary, the minimum acceptable being a 2 m^3 (6½ ft^3) outdoor enclosure with a shed-type shelter attached; the shelter will require a little warmth during the winter when these birds should remain inside. The outdoor floor should be of a sandy, gravel-like finish, clear in patches but planted with heathers, lavender, rosemary and such-like. A rear side can be covered with a strong growth of hops. Few perches are needed but a little roofed area outdoors is helpful to cover nesting boxes and provide a dry roosting place when weather is wet or chill winds blow.

Food
This is one of the hardest birds to keep in top condition over a full active life-time. When I know that a bird keeper is capable of moulting this bird into and out of colour year after year then I have real faith in his proficiency in bird management.

The diet must be rich yet not fattening. If it is over-rich the kidneys may suffer; if it lacks nutrition the eyes will begin to exhibit sticky, watering edges. If the situation is not rectified, the nictitating membrane will close, badly limiting the sight. Even at this latter stage one can 'bring back' these birds; the daily use of such vitamin supplements as Abidec (Parke, Davis) and a complete change of food to an improved diet can slowly set about an improvement. To take on the management of these birds if not fully capable of caring for them is not justifiable, and one embarking upon this move should firstly acquire the necessary experience or be in a position where he may call upon, and receive, continual advice from more experienced people.

Give a well-balanced insectivorous mixture, dampened with milk in preference to water, but serve in a dry or fine crumbly condition. To this add finely grated cheese, steamed and finely minced ox heart, liver, fish and roe; minced shrimps or prawns now and again are also acceptable. Add the usual chopped or minced dandelion or comfrey and fresh elderberries when available. A small number, say about three per bird per day, of mealworms, each cut into about six pieces and well mixed into the food, should be given plus a similar number of very young locusts, or moth or wood-ant larvae.

Unless hand-reared these birds will rarely take a paste, to which could be added an assortment of items, as and when one wishes. If they can be persuaded to take such a food as a supplement so much the better; give a small amount daily and use it to introduce variety.

If these birds are housed in pairs instead of singly, and one sees one bird suddenly attack another, keep the aggressor under observation. These birds are normally very placid, with a strong pair-bond, but they can become rather spiteful when the diet is unsuitable and they are being 'starved' of some requirement. If the bird makes a habit of attacking the other, particularly when it is about to be fed or whilst it is searching for something in the food dish, house the culprit on its own for a few weeks, offering a course of a reliable vitamin additive, perhaps in the drinking water, and taking the bath out of their enclosure until later in the day, when one is sure the drinking facilities have been made full use of. As an alternative one can add a vitamin preparation to the milk which moistens the softfood. This should never become a matter of habit, look upon a course of supplements as a remedy to be used whenever there is a strain upon the bird or something seems not quite right. The diet needs to be checked carefully if such behaviour does occur in one's stock.

Although the provision of insect life is important, the softfood is the basic ingredient for good health. It is no good offering only the insects which happen to be available to you, without fully investigating the diet in the wild and varying it accordingly. To give insect life in quantity under controlled conditions is not a good thing as, even when offering half-a-dozen different insects, one is not providing the equivalent of the wild and varied intake. Far better to ensure the provision of a well-balanced basic diet of man-made food and, more important, that it is being eaten daily by the bird. The insects included in the food then become an inducement to eat more of the basic mixture. When young are being fed, either by parents or if hand-reared by their keeper, always cut most insects into small pieces and mix them into the food, whether mealworms, moth larvae or locusts. Mealworms, of course, should always be offered in this state, never alive or whole.

Breeding
I must admit that I have never yet bred this bird to maturity under its natural parents though I have hand-reared it on numerous occasions. When I have had pairs in the past, something else always seemed to crop up which was more important than my pairs of pied flycatchers, or they had to share an aviary with non-breeding spare birds. They have got as far as laying eggs on a few rare occasions, having nested in a variety of box-type sites; the provision of these, facing in a number of directions, should be made early in the autumn in their breeding area and they should be left to weather outdoors over the winter. The birds seem to take to these more readily than to the freshly installed site.

Provide mosses, grasses, dead leaves and as many different kinds of animal hair as can be gathered: fine dog hair, horse hair cut into short lengths and a few feathers. These birds seem to vary widely in their choice of nesting material and some nests have contained totally different items from others, such as spiders' webs and lichen. It is better to give a little of each and observe which materials the individual birds select, then supply accordingly.

The foods used when hand-rearing birds can be so vastly different to that given by the parents, that one cannot really compare them. I have used wax-moth larvae and a paste of our own softbill mixture, with various additives such as calcium lactate, greenfoods and fish roe steamed and crumbled, the whole being moistened with milk. When the parents are rearing they use many caterpillars and other small soft-skinned insects. The safest diet for success is wax-moth larvae, wood-ant larvae, freshly hatched locusts and two dozen or so finely cut-up mealworms mixed into the basic food mixture four or five times a day; only the mealworms and large locusts need be cut up.

White-collared Flycatcher (*Ficedula albicollis*)[**]

General Description
This bird is very like the pied flycatcher except for a white rump, a full white collar around the neck and slightly more white upon the forehead. The song is slightly different, being softer and more liquid. The male is totally black and white during the breeding season whilst the female is grey, rather more so than a pied female, and shows, upon close inspection, a slight lightish-grey patch on the rump and around the neck where her mate is white. There is a little more white on the wing of the collared bird in both sexes but one needs to have examples of both species in front of one to compare such differences. The length is again 13 cm (5 in) or just a fraction below.

Habitat
Although overlapping the area of the pied flycatcher, this migrant is a more southerly race in general but has a similar habitat.

Aviary
Treat exactly as the pied flycatcher (page 174).

Food
Their dietary requirements are the same as that of the pied flycatcher (page 174).

Breeding
Treat as the pied flycatcher (page 175).

Azure Tit (*Parus cyanus*)

General Description
This tit is longer than the blue tit, measuring some 13 cm (5 in) in length. The extra length is not in the body but taken up by the slender tail which is blue, edged and tipped with white. The eye stripe is bluish-black and there is a small patch of the same colour in the centre of the breast; the crown, cheeks and under-body are white and the wings and mantle bluish-grey, but the wings have a brownish wash. The outer feathers are more blue and edged with white.

Habitat
This bird frequents the banks of streams and broad-leaved woods rather than dense forests, invariably keeping to the fringes in such places as willow thickets and birch groves.

Aviary
This is a very active and restless member of the tit family, used to fairly strong flight and it should therefore be provided with an aviary of more than the usual length, perhaps 3 m (9¾ ft) long by 1 m (3¼ ft) wide and 2 m (6½ ft) high. A small roofed section at each end should shelter hole-entrance-type nest boxes and these should be left in position all the year round. One side of the enclosure should have a thick creeper, such as hops, to encourage insects. The ground should be mainly gravel and sand, with possibly rockery plants for decorative purposes; include two cotoneasters (*Rosacea horizontalis*) as the berries will be appreciated in the autumn and the bushes will also serve the bulk of the perching needs. A further perch at each end beneath the sheltered sections will suit their requirements. Never house more than a pair together.

Food
A good insectivorous mixture should be given with grated cheese, two or three cut-up mealworms daily, steamed fish, or fish roe occasionally, mixed well into the food. Any small insects can be offered in limited

quantities; always bear in mind that the birds should be encouraged to eat plenty of softfood, so the livefood should be added as an attraction. Some of these birds peck at a slice of sweet apple while others seem to completely ignore it; a little could be minced in their food together with small quantities of greenfood. Peanuts will be taken in moderation as too will sunflower seeds. There is no need to hang a piece of suet or beef fat in their aviary because with a well-balanced diet, there should be no call for it.

Breeding
These birds should prove no more difficult to breed than the near relative, the blue tit. Log-type nest boxes can be used, either home-made ones or natural ones in the form of an old birch-tree bole, rotten and fallen, but having numerous old excavations by woodpeckers. The latter type can look quite attractive when an enclosure is planted naturally. For the first 5 days after young have hatched offer small soft-skinned insects; aphids are ideal but difficult to collect in bulk although a hop plant should provide many if one is planted in the aviary. One can give ant and moth larvae and increase the number of mealworms during this time to 50 cut up in the food. Give these night and morning. As the young get older so the insects can be increased in size and quantity. When the birds are older than 5 days, maggots can be given by the dishful; add a few drops of cod-liver oil, no more than ten drops to ¼ litre (½ pt) of maggots, and then powder them well with Vionate (Squibbs). Maggots in particular hold this powder very well if first moistened with a few drops of cod-liver or olive oil. All such oil should be used very sparingly; one is aiming only to cover the insect with the slightest film of oil to which additives can adhere, although the oil itself is beneficial.

The nesting materials are mainly moss and various kinds of hair; it is all so closely woven that it becomes almost like felt to the touch. The eggs are very similar to those of the blue tit, being white and spotted with dull red. The young also resemble their near kin in leaving the nest at about 18 days. With freshly hatched young, it is always advisable to use calcium lactate for dusting livefood.

Blue Tit (*Parus caeruleus*) *****‡

General Description
This is a remarkably well-coloured creature and measures 11.5 cm (4½ in) long. It is blue-capped and yellow-breasted, with a greenish mantle and blue wings and tail. It has white cheeks and blue legs and

a blue-black line through the eye to the nape, descending beneath the cheeks to join under the bill and cover the throat. It is very active.

Habitat
It frequents almost any park, garden, orchard or olive grove, almost always near human habitation, but in the main avoids coniferous areas.

Aviary
A 2 m (6½ ft) long aviary, by the same height and 1 m (3¼ ft) wide, as advised as the barest minimum for breeding birds, will certainly house a pair of blue tits. The vegetation growing in this enclosure should be as for the azure tit (page 177), as should the nest boxes which will often be used for roosting during inclement weather. Keep one pair strictly on their own; do not mix them.

Breeding
The biggest problem is not in getting these birds to nest, but in finding sufficient food to keep up to a dozen youngsters alive whilst they are being reared. Provide a box-type nesting site, either log or man-made; the hole would naturally be small but this matters little when the birds are housed on their own, as they should be. Nesting materials such as moss, fine grasses, wool, vegetable down, carpet sweepings, animal hair and fur, feathers, leaves, spiders' webs, even pieces of decorative wrapping, are often used. So many unexpected materials turn up in their nest boxes that they must be the most unfussy of birds in this respect. The nest is a warm felt-like cup surrounded by a wide variety of materials and contains white eggs marked with fine spots and smudges of dark brownish-red to purple-red.

The feeding of young should be as for the azure tit (page 178), a close relative, to which all this bird's needs may be likened.

Coal Tit (*Parus ater*)**

General Description
Of a similar length to the blue tit, 11.5 cm (4½ in), the coal tit is slimmer in appearance. It is black-capped with a white nape patch and cheeks and a black throat. The upper parts are slate-blue, the underparts white, the flanks smokey-buff, and the wings and tail are almost black. Like most of the tit family, it is an active little creature.

Habitat
Although it is more inclined than the blue tit towards coniferous areas it is still a frequent visitor to parks and large gardens. It nests in holes in banks, trees and walls.

Aviary
The minimum-sized aviary will adequately provide a comfortable residence for a pair of these birds. Fix log-type or box nest sites in position beneath a roofed portion of the enclosure. Plant out the aviary as for the azure tit (page 177) with perhaps the addition of a larch at each end situated not too far from the nest boxes; this can always be 'topped' and stunted to prevent it damaging the wire netting. House one pair only to an aviary.

Food
Provide a diet identical to that of the azure tit (page 177).

Breeding
These birds normally have fewer young than the blue tit, the usual maximum being about seven or eight. The rearing task is therefore slightly easier. This bird can be treated in a like manner to the azure tit (page 178). Nesting materials vary little if at all so provide moss, fine grasses and a variety of animal hair, plant down, feathers and spiders' webs.

Crested Tit (*Parus cristatus*)[†]

General Description
Crested, with long upturning feathers of black but white-flecked, this bird's upper parts are greyish-brown, with a rufous shade to the rump. The cheeks are white-edged, and a black collar covers the throat whilst a further black line runs back from the eye then curves round behind the ear coverts. It is a rufous buff on the flanks and the wings and tail are brown. The overall length is 11.5 cm (4½ in). It is very active, feeding treecreeper fashion and on ground.

Habitat
Frequenting coniferous woods, and alders if nearby, not usually too far from human habitation, it is to be found in Scotland and throughout Europe, including Scandinavia, as far as the Urals. It is not found in England.

Aviary
The standard minimum will suffice as a home for a pair of these birds. A small section of the roof should be covered at each end of their enclosure; beneath these site a number of next boxes. Since these birds often feed in a similar manner to the treecreeper, I have seen aviaries backed to advantage with vertical lengths of pine; these were split down the middle with a few branches left to protrude for perching. A few stunted pines in the aviary makes a natural and attractive setting. I would still suggest that one side be left for a climber such as the hop. I cannot emphasise too much the value of this in supplying aphids for the aviary inmates.

Food
Provide a diet as for the azure tit (page 177).

Breeding
Having watched many of these birds abroad, observed them rearing, hand-reared them and kept them over a long period, I have little hesitation in proclaiming them to be no more difficult to breed than any other member of the tit family so far dealt with. Treat in a manner identical to that suggested for the azure tit (page 177) for all feeding requirements. These birds will often eat from rowan berries during the autumn. Nestlings' livefood must be varied. The nest materials are grasses, moss, feathers, wool and fine hair.

Great Tit (*Parus major*) ****

General Description
White-cheeked, the great tit has a contrasting black crown and neck-line joining a bib and black throat, the black of which runs centrally down the bright yellow breast and belly. The back is yellowish-green, the rump and upper tail coverts are slatish-grey and the wings and tail are black edged with slate-blue. The length is 14 cm (5½ in). Its chief characteristic is its pugnacity.

Habitat
It is common almost everywhere within reach of human habitation, though more rare in thick conifer forest.

Aviary
Due to their pugnacious nature, these birds should never be kept in a

mixed collection. Even when an abundance of food is present, they invariably disgrace themselves. A standard minimum enclosure is large enough to breed this bird. Plant as for the azure tit (page 177) but include *Pyracantha coccinea*, the berries of which it will take towards the end of the year. A small area of the roof at each end of the aviary should be covered to give some shelter. Nest boxes should be left throughout the year, just cleaned in early spring by scraping; if other treatment is attempted these tits can take offence and quite easily turn away from the sites.

Food
Provide a diet as for the azure tit (page 177).

Breeding
They are a very common bird, yet only once have I attempted to breed them. I have hand-reared them often, kept them for observation purposes and even had liberated pairs nesting at my door. They should present few if any problems. Offer such materials as dry grasses, moss, rootlets, animal fur and hair, feathers and spiders' webs.

The number of young produced can be up to 15 but this number is rare; it usually is nearer eight or nine. Nevertheless one should be prepared. Food for rearing is as for the azure tit (page 178).

Long-tailed Tit (*Aegithalos caudatus*)**

General Description
Over half of the 14 cm (5½ in) length is accounted for by a long slender tail. At first glance one sees a black, white and pink bird, usually in a family group, flitting among tall hedgerows where it will be searching for insects or even berries during the winter months. The northern and eastern forms have clear white heads, whilst the western bird has a wide black stripe over the eye which joins the black of the mantle. The upper parts are dark with a pinkish wash, the rump and shoulders are rose-pink and the wings and tail have white outer feathers. The sexes are very similar, but with a number to compare, one can sex the western form by the white line running from forehead to nape in the male. If the white of the head is slightly greyish-tinged, fading towards the black edge, it is fairly safe to assume that the bird is female. The male has a very white, wide, distinct division on the crown. As with any such sexing, one can make an error through well-marked females or poorly marked males. Birds from the far north of the former USSR have slightly longer tails than others.

Habitat
The most widespread member of the tit family, this bird occurs across Eurasia, through China almost to the Himalayas; in the British Isles it is more common in the south than the north. It is a typical woodland-loving bird, a resident frequenting heavy bramble growth, dense hawthorn, young tree saplings where caterpillars abound and coppices rather than gardens. In the depth of winter, however they will at times venture in small family groups or flocks into more built-up areas.

Aviary
The standard minimum is just about adequate for a pair of long-tailed tits but an aviary longer by at least another metre (3¼ ft) is far better. Plant densely on two sides with honeysuckle and hops. Into these climbers, which should be kept separate at opposite sides of the enclosure, or at one end and one side, push a number of clumps of bracken, either dead or growing. This aviary should have a small roofed portion at each end and a frost-free enclosure attached. It need not be heated in any way but it would greatly assist the birds' condition to provide extra feeding time at night during the winter months.

Food
A nutritious softfood mixture will be needed for the basic diet; to this add grated cheese, steamed and minced ox heart, liver, fish and fish roe; sweet apple and small amounts of vegetables, such as dandelion and brassicas, can be minced and mixed in the food. Mealworms must be kept to a minimum and, whenever given should, be cut into small pieces then mixed well into the mixture. Encourage these birds to accept the Farlene (Glaxo) paste; it may first need to have some grated cheese mixed with it, so that they recognise the flavour of a familiar food. Small insects can be given daily but not too many unless the birds are breeding and rearing young.

Breeding
With the arrival of spring, check that the clusters of bracken are still in position in the climbing plants; these frequently help to form the foundation of a nest. The materials to offer are varied: as many spiders' webs as possible which can be collected on twigs bent to the shape of a tennis racquet frame, moss, lichen, feathers, animal hair and even small lengths of teased-out soft fibrous string. No nest will be constructed properly without the spiders' webs as this binds the domed structure together, not only during construction, but also when the birds decorate and weather the exterior with lichen. Many years ago, spiders' webs

were used in medicine to stop bleeding of wounds; the long-tailed tit seems aware of the moisture-barrier properties and covers the outside of the nest with this material, then camouflages it with a mass of lichen.

I have hand-reared these birds on numerous occasions, using waxmoth larvae, ants' 'eggs', wasp and bee grubs and mealworms and a softfood paste. A number of these soft-skinned insects, with mealworms cut small, should be adequate for a pair to rear young satisfactorily and I have seen it done.

Marsh Tit (*Parus palustris*)[†]

General Description
The crown as far as the nape is a glossy black and this extends below the bill to form a bib. The upper mantle and back are greyish-brown washed with olive; this fades out towards the rump and the wings and tail are also this grey-brown shade. The lower belly is dull white, whilst the sides, breast and flanks are pale buff. This bird is very like the willow tit but without the sooty matt-black crown. Juveniles, however, can easily be confused with one another. All have white cheeks and the sexes are alike. The length is 11.5 cm (4½ in).

Habitat
Woods, damp localities and orchards all seem to attract this bird but marsh seems to have no special appeal for it. Near broad-leaved trees and steep banked streams one can often hear its 'schup-p-schup-p-schup-p'.

Aviary
The standard minimum will again provide adequate space for this bird for breeding. Give hollow log-type sites, either nest boxes or old birch boles, which may have been found rotting on their sides in the woods; these, if stood erect in an aviary corner, form natural sites and an old woodpecker hole may soon be accepted. The aviary should have a roofed section at each end for cover and, although a shed-type shelter is not vital, the provision of one allows extra feeding time to be given during short winter days.

Food
Feed as the azure tit (page 177).

Breeding
Twigs, grass, moss, wool, hair and feathers will often, though not always, be used. It is normal for this bird to nest low in the wild but in an aviary it may quite often choose a high nest box. These, with the alternative hollow logs, can be at various heights. Treat in all other respects as for the azure tit (page 178).

Willow Tit (*Parus montanus*)[†]

General Description
The description of the marsh tit also fits this bird except that the crown in this variety is a duller black. The bib, too, is a little larger and tends to fade out rather than end abruptly as in the marsh tit. When handling it, one can discern a pale buff patch on the secondaries of the wing. The length is 11.5 cm (4½ in).

Habitat
This is identical to that of the marsh tit (page 184).

Aviary
Treat as for the marsh tit (page 184), but since this bird very often excavates its own nest hole in rotting wood, tree boles of a soft crumbly consistency should be provided in addition to the usual boxes and hollow logs.

Food
As for the azure tit (page 177).

Breeding
Since this bird frequently excavates the nest cavity, the inner part is covered with wood chippings. Rarely in such a case does this bird build a nest; instead it shapes these fragments to suit itself, then lines the hollow with animal hair, vegetable down and sometimes feathers. There are occasions when it has been known to use old woodpecker excavations and other cavity sites and then quite often a little more nesting material is used. In other aspects we can treat this bird in the same manner as the marsh tit (above).

Bearded Reedling or Bearded Tit (*Panurus biarmicus*)****

General Description
The head of the male is a clear blue-grey and the 'moustaches' start at the lores, reach the eye, then turn in an abrupt and elongated downward sweep to each side of the upper breast. The back is a rich chestnut-fawn, black beneath the tail coverts, whilst the tail itself is a rusty-red with the outer feathers edged and tipped with grey. The throat and upper breast are white with a greyish wash and the flanks are a pinkish-rust. The female lacks the blue-grey on the head, which is a pale buff, and she has no black beneath the tail. Both sexes have a long tail, well over 8.5 cm (3¼ in). The juvenile resembles the female but back and crown are thinly striated with black. They can be sexed at this stage: the male has an orange bill, whilst the female's bill is bone-coloured; the male also has a small amount of black in the ventral area.

Habitat
Marshes and fen areas where large reed beds exist are favourite haunts. It is the seed of this common reed which forms the bulk of the food intake during winters in the wild; at other times aquatic insects are the source of the food supply.

Aviary
To breed this species, it is recommended that an aviary for each pair be at least 3 or 4 m^2 (9¾ or 13 ft^2) and 2 m (6½ ft) high. One secluded corner should then be set up to resemble the birds' wild habitat: a small pond, even if only 1 m^2 (3¼ ft^2), in which can be grown a dense mass of reeds, will suffice and long grasses should surround this area. Although it is not essential, I would suggest that some form of evergreen bush be planted very near and slightly overhanging the reed bed; this is for roosting purposes and also to shield any nesting site from heavy rainfall. Winter quarters should be attached to this aviary and, although winter artificial heat need not be supplied, the extra hours of feeding from the use of electric lighting will prove beneficial. The birds may have free use of the outdoor aviary at all times.

Food
The basic insectivorous mixture should have minced or grated cheese, minced steamed ox heart and liver, fish and fish roe, all mixed well into it. A daily quota of mealworms can be given but these should be cut into small pieces at all times and well stirred into the food; live wax-moth larvae, ant larvae and day-old locusts can all be given sparingly in the

food supply. No fruit need be included but some vegetable matter must be supplied; if hops are grown on a rear wire section of their aviary, they will provide a host of aphids which will be greedily taken. Many aquatic insects are high in calcium content, so if using maggots, remember to powder them well with Vionate (Squibbs) and calcium lactate; this will help to replace items from their natural diet which have been lost.

Breeding
The first breeding of these birds was from a pair that I passed to a friend; the following year I also succeeded but, whereas his pair had nested naturally in a reed bed, mine disregarded the offered natural sites and used a half-open-fronted nest box about 1.5 m (4⅞ ft) from the ground. So much for all my well intended efforts in reed planting! The rearing was not difficult as they found much in the way of insect life in the aviary, which contained a wide variety of plants and shrubs in addition to the reeds and dense undergrowth.

Whilst rearing is in progress offer all the insects possible, including ant and moth larvae, small locusts and caterpillars, which can all be given alive. Mealworms, as always, must be cut into small pieces and mixed well into the softfood; this guarantees that a certain amount of the softfood is consumed. The free use of mealworms alone can result in a bird becoming so addicted to them that it will eat little else and must then be 'meated off' before being returned to a well-balanced diet. If maggots are to be offered as a rearing food, hold them back until after the 5-day stage; even then they should be treated with the additives previously mentioned. Before this 'risk period' is over, use only soft-skinned insects.

Wren (*Troglodytes troglodytes*)[†*****‡]

General Description
This little creature of 9.5 cm (3¾ in) long is a very popular and well-known species. Mere brown it may be, but the feather pattern when studied is very attractive. The barred plumage, rich chestnut-brown shades with darker markings, the short, rather stumpy, cocked-up tail of rufous hue, the blurred flight and loud song endear it to all who know it.

Habitat
This bird ranges almost anywhere, from mountain to moorland, town park to country garden. The domed nest may be found in places as varied

as an ivy-covered tree, a garden shed or may be built into a garden birch-broom or the pocket of an old gardening jacket hanging in a shed.

Aviary
Certainly the standard minimum-sized enclosure will house a pair of these birds. Plant two sides with such creepers as honeysuckle or hops; hops may not look quite so attractive but they do help with the natural food supply of aphids. Give a small roofed area at each end of the enclosure. A shed would certainly be appreciated in cold winters but it is not vital if plenty of warm roosting sites are allowed to develop. Nest boxes could be sited amid the creepers, which will act as a wind break.

Food
The usual basic food should be supplied, with grated cheese added. Give three mealworms per bird cut up small in the daily food. If there is a danger of one bird getting too many, provide two separate feeding areas and they will very soon get to know their purpose. A little soft fruit, such as apple or pear, and some vegetable greenstuff, should be minced finely and given in the food. Insects such as moth and ant larvae, both storable in freezers, can be given all the year round; enough to keep them at the dish should be supplied. Softfood will always adhere to cut-up insects such as mealworms, and the live softer-skinned creatures' movements in the food will attract birds to it.

If the birds do not seem to be reaching breeding condition in time, or if you wish to make certain of their doing so, add extra animal protein in the form of ox heart, liver and perhaps fish roe to their diet. I like to do this as a matter of course with many species and I honestly believe that this is what has helped me to breed so many varieties over the years. The song of a male, or the absence of it, will soon let you know how your method of feeding is affecting the birds' general health.

Breeding
Nest boxes or natural sites may be used. At times two or three nests made by the male are ready for a female's inspection. Once a nest has been approved by her she sets about lining it; the materials one should offer are very fine short pieces of grasses, leaves, moss, animal hair and feathers. The birds will appreciate a wide variety of sites. Although the rule is one pair of birds to an aviary, any resulting young of this species can stay with the parents until early the following year. It is most likely that all will sleep together in an old nest that becomes the common roost and this is quite normal during cold weather.

Provide ample supplies of aphids, wax-moth larvae and cut-up mealworms in the softfood and even wasp or bee grubs if possible from the day the young hatch; since they are native to the British Isles they are fairly coarse feeders; earwigs and many other seemingly unlikely wren foods will be taken at all times.

10

Thrushes, 'Chats', Redstarts, Bluethroats, Nightingales, Siberian Rubythroat, Red-flanked Bluetail, Robin, Siberian Blue Robin, Wheatears and Accentors

THIS IS A LARGE GROUP of birds to describe in one chapter and diets do vary greatly in some instances, yet all are fairly coarse feeders. Not all, it must be emphasised, will survive on an identical diet: in some cases the basic food may hold very similar ingredients, but whereas, for example, thrushes prefer a fairly coarsely ground mixture, such birds as bluethroat, dunnock or red-flanked bluetail would all prefer a finely ground basic food.

A ripe apple, for example, with the skin broken, can be thrown into a thrush family's compartment and will be consumed forthwith; such is not the case with all the birds covered here and fruit should in general be either grated or minced; the moisture in food mixtures should come from pulped apple, pear, fig or other suitable fresh fruit. Diets for various birds may well remain similar but the preparation can be very different.

When not feeding young, all the birds of this group may have their livefood cut to a minimum as long as the basic softfood is adequately nourishing and they are taking such food readily. A slight increase may be made in the daily intake of livefood during the moult and again during the partial spring moult, when not only is there a plumage change of sorts, but the breeding condition is gradually being built up. Vast quantities of insect life, however, should never be necessary if the birds are wintered and fed correctly. The general management of stock will be reflected in the exhibited condition of the birds.

When considering coarse-feeding species of birds, one should be careful not to confuse coarse feeding with the provision of a dull boring diet. This can never be described as good husbandry; this type of diet may well keep birds healthy for a short life-span – a matter of months or even a year or so – but rarely will a bird keep in faultless condition or be capable of rearing full clutches, let alone of fertilising the eggs. One cannot expect the diet for a caged bird to fulfil the requirements of one that has a large enclosure to exercise in, just as the calorific food intake of a sedentary worker would be insufficient for a manual labourer. We give here the amount of food needed by a bird living in an aviary; a caged specimen on a similar diet could become fat, lethargic and probably develop an internal malfunction of one kind or another with eventual fatal results. Daily exercise over a large area promotes health and general fitness when taken in conjunction with a sensible diet.

The coarse feeders' main requirement is for a well-balanced protein–vitamin–mineral food form. Always give fresh vegetable matter, fruit and animal protein as well as farinaceous meals; it is a simple matter to include these if grated or minced finely. Forcing birds to take common earthworms, when they are not normally a part of their diet is unwise; bacteria can very easily be introduced in this manner. One bird may indeed have a built-in resistance to eating such items but another may well become infected with a serious disease through ingesting such organisms. Some vegetable growth, such as comfrey, is very high in mineral content and such foods should always be available. If there is a danger of mineral deficiency include a seaweed meal in the food mixture; if used as an additive of up to 5 per cent of the basic food any seaweed should supply adequate minerals and trace elements. During the autumn, when wild berries are at their most plentiful, collect elderberries and rowan fruit. These will be eaten greedily whilst supplies last and such fruit can easily be frozen for later use and even for continual feeding throughout the winter.

Wheatears, like redstarts, nightingales and a few other insectivorous species, can live together in pairs quite amicably in an indoor enclosure, only to fight when placed into a breeding pen outdoors in the spring. The introduction of such birds is recommended by using side-by-side enclosures with an adjoining door; when it is considered that peak condition has been reached by both birds and a response is observed, they can be 'paired' in their new joint enclosure.

Blackbird (*Turdus merula*)****‡

General Description
The male is totally black except for the bill, which is yellow for most of the year and changes to deep orange with the arrival of breeding condition. The length is 25 cm (10¼ in). The female is identical in size but has dark brown, verging on black, plumage. Young of the year are very rufous on the upper chest, dark like the female but found upon close inspection to be densely striated. Partially albinistic specimens are common and this abnormality varies in extent from a feather or two or evenly or unevenly marked white patches on head or body, to the entire wings or tail. The bills of female and young are a bone-yellow.

Habitat
The blackbird is a common garden bird, frequenting orchards, forests and most woodland or agricultural areas.

Aviary
I would recommend an aviary of 3 m (9¾ ft) in length, 2 m (6½ ft) high and at least 1 m (3¼ ft) wide for a pair of these birds. Beneath the perches it will be found best to add peat which may be turned over when soiled, or shingle so that hosing or rain will remove any accumulation of droppings. Certainly a fairly large area should be left clear. In other areas where there is less chance of vegetation becoming unsightly through soiling, the inclusion of evergreens or even flowering shrubs will both provide cover and add to the attractiveness of the enclosure. The structure needs only very little shelter which should consist of a small section of roof at each end being covered, and perhaps a board or two of timber below this to prevent driving rain causing the inmates discomfort. Perches should be kept to a minimum, with one at each end beneath the roofed section for roosting and two lower if no natural growth is available.

Food
Provide a coarsely ground insectivorous mixture. If on the fine side a little honey can be mixed with it. Include grated cheese, minced green and root vegetables, grated fruit or a few soaked household currants; three cut-up mealworms per bird daily stirred into the food will cover their livefood requirements during out of breeding season periods. I have always used steamed and minced ox heart and liver and also steamed fish when available; there is no need for large amounts of such foods and two teaspoonsful per pair is adequate when mixed well into the basic

food. Fresh ripe soft fruit should always be available; I have always found many greengrocers most helpful in providing over-ripe fruit if one's purpose is explained to them.

Breeding

Where adequate growing shrubbery exists to provide suitable nesting sites one rarely need provide additional sites. Some birds like a cluster of branches surrounded by evergreen and others need nothing. I have usually offered a choice, natural site and half-open-fronted box-type sites which were about 25 cm^3 (10 in^3), with a small board covering the lower half of the front. At other times they have used a box of the same area but only a third as deep and with no roof. Materials required will be a variety of coarse and fine grasses, dead leaves and weed stems; mud should be available for the lining of the inside of the outer shell; this can be from a small pond in their enclosure or a bowl holding mud and placed near their bathing facilities to keep it moist. The final lining will usually consist of fine dry grasses but depends entirely upon the native habitat of the birds; those in the far north use such materials as pine needles, whilst I have found some nests in Italy lined almost completely with leaves.

With the arrival of young increase the number of mealworms, cut up and mixed into the food. On the first day of hatching cut a dozen or so into small pieces and add to the food; the next day this needs to be doubled and again on the fourth day. On the second day after hatching, give such items as young locusts, moth larvae, wood-ant larvae and even a few well-cleaned maggots treated with a few drops of cod-liver oil (three or four drops to a 60 gm (2 oz) container); these additional insects should be in a separate dish so that one can tell at a glance how the foods are being taken. These supplies of live insects will have to be increased according to the number of growing young and their food consumption. As the birds increase in size, larger insects can be given; fully grown locusts will be broken up by parent birds and the softfood containing cut-up mealworms will also be passed on. Locusts are good value for money when well fed and an excellent rearing medium.

Occasionally, prior to pairing, one bird can be a little too aggressive. Endeavour to learn which is less advanced and increase its daily mealworm quota or offer a vitamin additive in its drinking water before placing them together. Never house more than one pair in the same enclosure and avoid having pairs even next door to each other. Keep a totally different species between their aviaries if using rows of flights.

Fieldfare (*Turdus pilaris*)***

General Description
This bird resembles a mistle thrush in size, being just over 26 cm (10¼ in) and thick-set. It is very aggressive and only amicable individuals should be kept together since much damage can otherwise result. The sexes are very similar, with a bluish-grey head and rump, very noticeable white under-wings, a rich chestnut back and wings slightly browner; the breast is golden-brown heavily spotted with black and the underparts are buffish.

Habitat
A bird of Scandinavia, it is found in open coniferous forests and areas of broad-leaved trees. During the winter it travels south to wherever it can find wild berries, on which it survives during these hard months, or agricultural ground.

Aviary
As for the blackbird (page 192).

Food
As for the blackbird (page 192).

Breeding
As for the blackbird (page 193).

Ring Ouzel (*Turdus torquatus*)**

General Description
At 24 cm (9½ in) long, this bird resembles the blackbird in colour and size, but the large almost half-moon shaped gorget will immediately identify the male. Upon closer inspection one will also find that he has paler wings with a grey edging to them. The female also shows this gorget but it is far less conspicuous. She is much browner in appearance with more conspicuous light edging to the wings. The song is wild and beautiful, perhaps not so much so as that of the blackbird but louder and, when heard at night or in a quiet rocky area, the clear pipe-like notes are very attractive.

Habitat
Mountainous rock-covered areas and wild moorland suit this bird very well. It will nest among the deep heather on the steep side of some ravine, beneath fallen branches or a growing bush, in a hollow beneath jutting rocks or, occasionally, low in tree or shrub.

Aviary
As for the blackbird (page 192).

Food
As for the blackbird (page 192).

Breeding
As for the blackbird (page 193) but add moss to the nesting requirements and cut down on the quantity of mud. There is no lining of mud in the ring ouzel's nest but the base is often 'cemented' to some solid foundation with it.

Redwing (*Turdus iliacus*) ****

General Description
Although the smallest of the thrushes common in the British Isles at 21 cm (8¼ in) long, it is by no means unattractive. The conspicuous white eye stripe and the deep chestnut-red flanks quickly distinguish it from other thrushes, though the spotted throat and breast are reminiscent of the song thrush.

Habitat
This is basically a Scandinavian birch forest and scrubland species, although it now nests over a slightly wider range, wintering in gardens and agricultural areas throughout Europe.

Aviary
This is the only member of the thrush family that will, in my experience, breed on the colony system but I am not advocating it, although very little bickering takes place in a large aviary containing a few pairs. For one pair provide an enclosure as recommended for the blackbird (page 192). Should one wish to house a family together, original parents and offspring, such an enclosure would have to be extended correspondingly.

Food
As for the blackbird (page 192).

Breeding
Supply all materials as for the blackbird (page 193) but add mosses and lichen, together with some rotting crumbling wood pulp; a decayed birch will normally prove ample. This bird will often use this pulp with mud, to line the outer cup frame; fine dry grasses will then form the inner lining. All other needs are as for the blackbird (page 193).

Mistle Thrush (*Turdus viscivorus*) †*****

General Description
Measuring at least 27 cm (10½ in) long, this bird is far more grey than the song thrush. The breast spots are larger and the face noticeably paler and the outer tail feathers are white tipped. The song, although very loud, lacks the musical qualities of both the blackbird and song thrush.

Habitat
This bird frequents well-wooded country but where trees, both coniferous and broad-leaved, are not too dense. It often nests in trees adjoining agricultural land or parks; even large roomy gardens or orchards which will provide most of its needs.

Aviary
The enclosure as recommended for the blackbird (page 192) would be considered the minimum size for this bird if it is hoped to breed from it.

Food
As for the blackbird (page 192).

Breeding
As for the blackbird (page 193).

Rock Thrush (*Monticola saxatilis*) ****

General Description
Here we have a most strikingly attractive bird. The male is a wonderful sight when in breeding plumage, with a Cambridge-blue head, neck and throat, orange-chestnut breast with a crescent of this hue on each

feather, a conspicuous white back patch seen more in flight than when at rest, almost black wings, and a rich-chestnut tail. The female is much browner and is finely speckled with none of the male's brilliant hues. Young resemble the hen in every way and are not easily sexed until the adult moult in the spring; on close observation one can tell the male by his white back patch which shows very slightly once out of nest feather. The length of this bird is 19 cm (7½ in).

Habitat
It inhabits the mountainous areas of Europe but is rarely seen because of its shyness and habit of 'freezing' into a watching position when spotting an intruder. It nests in a rock cavity.

Aviary
As for the blackbird (page 192).

Food
As for the blackbird (page 192).

Breeding
It readily accepts a half-open-fronted nest box. In all other respects treat as a blackbird (page 193).

Song Thrush (*Turdus philomelos*)[†*****‡]

General Description
The commonest of British thrushes, it is much appreciated for its song. The overall appearance is of a warm, slightly olive-tinted brown; the underparts are white, the breast is covered with dark brown spots and there is a slight orange tinge to the breast and flanks. The back always seems very nightingale-like. The length is 23 cm (9 in). The female is almost identical to the male though her breast spots may be larger and recede lower on the belly, whereas they stop short of the lower belly in the male; he carries more white on the ventral area.

Habitat
Common everywhere in field, orchard, sparse woodland, coniferous or broad-leaved tree forests and town or country garden, it is often chased away by blackbirds, who also like to be near human habitation and the food that this proximity provides.

Aviary
As for the blackbird (page 192).

Food
As for the blackbird (page 192).

Breeding
As for the blackbird (page 193), but include wood pulp with nesting materials.

Whinchat (*Saxicola rubetra*)****

General Description
To say that the plumage is made up of various browns and black hardly seems to do this bird justice, for it is a remarkably attractive chat in breeding plumage. The male has a black cheek patch, a very conspicuous white eye stripe and a white wing bar with almost black shoulder patch and flights. The rich brown of the back and crown is evenly marked with a brown so dark it is almost black. The throat at the edge of the black cheek is white and then, from here to the belly he is a warm orange-tan fading to white at the vent. The female is more brown all over and, after the autumn moult, both look very much alike. The young are similar to this eclipse plumage.

Habitat
Open ground and moorland provide favoured haunts where it nests beneath an overhanging tuft of grass.

Aviary
As long as one does not have to enter the enclosure to feed or water the birds, a 3 m (9¾ ft) by 1 m (3¼ ft) by 2 m (6½ ft) high aviary would provide ample space for a breeding pair of these birds. Allow grass to cover half of the floor area; the remainder should be of shingle and peat with heathers and perhaps an evergreen or two. Winter quarters should be attached unless the birds are to be caught up in the autumn for transfer to warmer indoor flights. This bird can be allowed to leave such an indoor flight and use the outdoor aviary at all times, but ensure the food supply is indoors so that it is encouraged to enter for food and roost inside.

Food
As for the blackbird (page 192) but use a more finely ground basic food. Fruit can be grated fig only, with a supply of elderberries when they are ripe on the trees. Give grated or minced vegetable, brassicas, comfrey and dandelion, plenty of grated cheese and steamed ox heart and liver. With insects kept to a minimum at all times other than when rearing is in progress, extra feeding time during winter is essential.

Breeding
Provide grasses, moss, and animal hair for nesting materials. As well as ensuring that natural grass sites form, hang one or two half-open-fronted type boxes at the rear. A biscuit tin loosely packed with dry bracken and grass can be erected on a post to stand about 1 m (3¼ ft) from the ground; this type of site is often accepted in preference to others. All advice on rearing is as for the blackbird (page 193), except that small locusts rather than fully grown ones should be given at all times.

Stonechat (*Saxicola torquata*)[1****]

General Description
In breeding plumage the male is predominantly black and chestnut-brown. Though the white side neck patches and white in the wings are exploited in the mating display, so much white remains hidden at other times that the large areas of it bared at this time are almost unbelievable. The head, throat and back are black and the chest is chestnut fading downwards to buff. The female is browner on the head and mantle at all times and the young resemble her. The length is 12.5 cm (5 in).

Habitat
Gorse bushes seem the favourite of this bird and, even if only one or two exist in otherwise open moorland, the male will invariably choose one and keep to a particular perch for singing. He has a song flight, too, when he rises almost vertically, then rises and falls over a distance of 3 or 4 m (9¾ or 13 ft) delivering his courtship song.

Aviary
Because of this bird's shyness when nesting, try to give a large enclosure of 3 m (9¾ ft) long, 2 m (6½ ft) high and 2 or 3 m (6½ or 9¾ ft) wide. If planting naturally with gorse, procure small bushes without damaging the tap-root; the larger ones often die when moved. Heather

is another useful plant for this species and, if one can manage a mixture of the two, covering half the floor area, with sand and shingle for the rest, these birds will settle in very quickly. For winter quarters there should be a frost-free enclosure attached where they can retire at night or during the colder weather. Extra feeding time is beneficial although this is a very late-roosting species.

Food
As for the whinchat (page 199).

Breeding
As for the whinchat (page 199).

Black Redstart (*Phoenicurus ochruros*)[1*****]

General Description
The male in breeding plumage is blackish all over except for the rufous tail and white wing patch; the mantle is slightly more grey but the face and breast are true black. Some males do not reach this full adult plumage for 2 years; I have seen many breeding in what can only be called their juvenile plumage. The female is dark grey all over with the merest lightish-grey showing in lieu of the white wing patch. The young resemble the female. The length is 14 cm (5½ in). The song is rather like the noise of a small whistle when drenched with saliva, a kind of bubbly warble which is easily identifiable.

Habitat
The habitat of this migrant varies greatly with the country of residence, ranging from ruined houses and large desolated buildings, where structural steelwork or holes in the wall can provide nesting sites to rocky hill country where nearby villages and farm outbuildings can cater for this preference for rock or wall cavities.

Aviary
The most successful breedings have occurred in aviaries of at least 2 m (6½ ft) long; these need a width of at least 1 m (3¼ ft) and a height of 2 m (6½ ft). Since a pair of these birds rarely if ever squabble, unlike its close relative the redstart, one can leave them together without fear of losses. A shed-type shelter should adjoin their aviary and they can be allowed to use both indoor and outdoor quarters during the colder months. Extra lighting for longer feeding hours is an asset in keeping

them in top condition and advancing it towards the spring when they should breed. Where no such shed exists, place them in a flight as large as possible or in an unheated bird room: as long as they remain in frost-free conditions and are fed correctly they will stay fit throughout the winter.

Food

I like to keep these birds almost entirely on a good quality insectivorous mixture. When they do have mealworms, each insect is cut into small pieces and mixed well into the food; the only live insects I provide are wax-moths and their larvae. They receive a generous ration of grated cheese, steamed ox heart and liver which has been finely minced, vegetable matter and a little fruit, which has also been minced; now and again I give carrot and the roots of dandelions. Of course one can keep them on a far more austere diet but it is the variety which makes all the difference when breeding is top-most in the bird-keeper's mind. I honestly believe that this extra attention to detail is responsible for many of my own first breedings and other successes.

Breeding

I could not even guess at how many pairs of these birds I have kept since a child, and I have never known a case of severe fighting between a true pair. One can, if not careful, mistake a young male for a female, but a true pair have never exhibited any bad feeling towards one another when in my care, and I have rarely been without them. This amity usually simplifies keeping, pairing and breeding in general but I have known two hens to go through the complete breeding ritual, one nesting, laying and brooding the useless clutch whilst the other female would sit on guard, frequently calling off the brooding hen upon the approach of human beings; this well emphasises the importance of correctly sexing one's birds if they are in juvenile plumage.

Provide half-open-fronted-type nest boxes, some need not even have the roofed section. The black redstart is very easily accommodated; I have had hens which used a shelf and built alongside a vertical post. The materials they will need for nesting are: dry grasses, moss, fibrous plant skin, which they will take from dead vegetation such as honeysuckle and iris, animal hair, feathers and small particles of wool. With the hatching of young do not over-feed with a sudden glut of livefood; a young bird is quickly satisfied at this stage and can easily cease to gape and this, to a parent bird, is a dead youngster and, as such, may be thrown out of the nest. It is far better to allow double the normal quantity of cut-up mealworms in the food mixture and just a small quantity of wood-ant or wax-moth larvae in the initial first hatching period; they can be increased gradually from then on.

Redstart (*Phoenicurus phoenicurus*)[†*****‡]

General Description
A well-coloured male in breeding plumage is one of our more exotically coloured birds; he has a grey-blue mantle and crown, white forehead, black throat, brilliant chestnut breast and tail and darker wings. The contrast where the black cheeks meet white, grey-blue and chestnut make him a most handsome specimen. The length is 14 cm (5½ in). The female and the young are dull brown but retain the red-chestnut tail. The youngsters are easily identifiable by their spotted general appearance.

Habitat
Oaks and beeches with nesting holes are great favourites, but this bird is found equally in coniferous and broad-leaved forest areas, and even open scrubland can attract it at breeding time. An old army petrol can lying on its side on a shooting range once proved to be the nesting site of a pair, and they reared a brood of five successfully. Well-matured orchards may well hold a pair or two. On its northern limits, it has only to arrive a week later than usual and starlings or great tits will have taken over its usual sites.

Aviary
On rare occasions these birds have been known to breed in very small enclosures, even a large flight cage if they have not been moved outdoors soon enough from their winter housing. This is hardly the ideal situation in which to breed them, however, and they are unlikely to produce full nests of strong youngsters. The problem with these birds is pair acceptance, and I would advise the would-be breeder of this species to use two of the standard minimum-sized enclosures side by side with an adjoining door. Each of these aviaries should hold suitable nesting boxes, those with an entrance hole being most acceptable. The rear of each enclosure could have hop plants climbing the wire as an attraction to aphids, thus providing a ready supply of natural food. The floor space can be planted with a variety of flowering shrubs and evergreens. A frost-free, shed-type shelter should join on to this outdoor breeding pen, or suitable indoor accommodation made available for wintering these birds separately.

Food
As for the black redstart (page 201).

Breeding
Some birds are always a problem to pair and the redstart can be one of these. To understand this one must consider their natural method of pairing. In the wild a male will return to a breeding area and take up a territory; he will sing and thus attract a female to this area and then hover butterfly-like in front of a chosen nest site; the female, if in the appropriate condition, will follow him. Should she not do this, and remain aloof, he will invariably chase her from his territory. This procedure will carry on until the 'right' hen accepts both him and the nest site; she will enter and thereafter there will be no squabbling. We should bear this in mind when pairing our birds. We can hardly keep unlimited numbers of females to try with each male but, by housing them in adjoining aviaries, we will see when the female reaches breeding condition by observing her behaviour when the male tries to attract her to a nest site in his enclosure. As he hovers in front of the nest hole of a box near the dividing wire of their two aviaries, she will endeavour to investigate the site by flying to the nearest point of the division. Let this carry on for a day – no longer – and, whilst the male is going through his display, open the adjoining door; this can be managed with a length of string so that entry into the enclosure is not necessary. Keep the pair under observation and, if he attempts to attack her and drive her off, separate the birds for a few more days. I have used this method with a hen on each side of a male and never yet known it to fail. Under the normal method of introducing a pair, many birds have been lost, both males and females, depending upon which bird was more dominant at the time.

When a very old friend of mine of days gone by, Mr Reggie Tout, was attempting to breed a hybrid from the robin (*Erithacus rubecula*) and the redstart, I fostered eggs of each species under the other. The young were reared satisfactorily but, as I feared, no hybrids resulted; the mating behavioural patterns are so different that in neither case did it trigger off the required response. We liberty-bred redstarts also by this method, giving eggs from an aviary pair to wild robins for rearing, the robins having nested in an enclosure which had been left open and empty over winter. Such experiments are interesting but it is difficult to generalise from the information gained when foster parents are used. The risk of losing redstart pairs to predators when liberated in a suburban garden was too great and whether they, too, would have managed to rear the young on the livefood available in such an area was one question which remained unanswered.

For information on rearing the young redstarts when hatched, refer to the black redstart (page 201) and treat accordingly.

Red-spotted Bluethroat (*Luscinia svecica svecica*)**** and White-spotted Bluethroat (*L.s. cyanecula*)**

General Description
As they are so closely related we will deal with these two birds under the same heading. They are both very robin-like in movement. In the male the mantle and back are a rich dark brown and the tail is rufous. There is a noticeable pale eye stripe and the throat is blue; this extends to the upper breast where it ends with a darker hue. A thin, almost white, line here separates this blue from a rich reddish-chestnut crescent across the width of the chest; beneath this he is pale buffish, fading to almost white at the vent. The one difference clearly visible in the throat markings of adult males is the red spot in one and the white spot in the other. The females, as adults, can also be identified in this same manner; the red-spotted variety clearly exhibits a faded chestnut spot above a far paler blue upper chest patch and the white-spotted female shows little if any blue, with only a whitish spot to her throat. The moult in the autumn brings a much paler plumage to the red-spotted male: the blue recedes lower, the throat barely shows a red spot and even the chestnut band is duller. The male of the white-spotted race fades correspondingly and is slightly paler still. The young are heavily spotted, rather like robins, redstarts or nightingales, and, like these birds, when they leave the nest, they invariably go to ground and can hardly fly. The length is 14 cm (5½ in).

Habitat
The mountain bogs of northern Europe, Scandinavia and the former USSR are the usual breeding areas of the red-spotted race. The white-spotted bird hails from central and southern Europe. Although still preferring the same swampy ground and stubbly vegetation, the bird from more southerly regions also frequents drier mountainous areas.

Aviary
A minimum of 3 m (9¾ ft) long by 2 m (6½ft) high and 1 m (3¾ ft) wide does not really do justice to these birds, they are seen at their best in a largish enclosure, perhaps rather boggy from the over-flow of the pond. Try planting some evergreens and a wide variety of flowering shrubs. Even when hand-reared, they are a little on the skulking side,

sometimes confiding enough to take food from the hand, but just as frequently disappearing into the undergrowth. Winter quarters should be attached, to avoid chasing and having to catch up these birds annually. Provide extra lighting during the shorter days but heat is not required when allowing this additional feeding time. They may use an outdoor aviary daily just so long as they have the warmer shelter to retire to when needed. This is another bird which will benefit from having hops growing on one side of the enclosure; hunting for insects is good for them and ensures adequate exercise being taken by a bird which can otherwise become obese.

Food
Keep mealworm consumption to a minimum and encourage bulk feeding on an insectivorous mixture. This bird can easily become sore-footed through too many mealworms in its daily intake; feed as for the black redstart (page 201). Bluethroats are very fond of elderberries during the moult.

Breeding
Although nesting on or very near the ground when at liberty, this bird, in an aviary, freely takes to a half-open-fronted box or an easily assembled site, such as a biscuit tin erected up to 1 m (3¼ ft) from the ground on a mouse-proof post; this tin should be packed loosely with dry bracken and grasses. With the tin positioned on one side so that the open top is facing away from full sunlight, it becomes fully weather-proof. I have used this method many times for nest sites and in fact the white-spotted bluethroat was bred for the first time in this very manner, not in my aviaries on that occasion, although from birds which I had passed on. Nesting materials should include such items as dry grasses, ranging from fine to coarse, moss, animal hair and vegetable down.

When fortunate enough to have young hatch do not be too liberal with mealworms and always cut them up small and mix well into the food. The larvae of the wax-moth are far less addictive and safer to feed in bulk. When rearing on wood-ant larvae, gradually decrease the supply as the young get older, increasing moth-larvae and young-locust supplies daily. I find that the young more readily take to an insectivorous mixture containing locusts and moth larvae rather than mealworms or ant larvae. When mealworms are cut up small in softfood as advised, it will be seen that many parent birds will take gradually increasing amounts of softfood to the young, the earlier smaller quantities having adhered to the pieces of mealworm.

Nightingale (*Luscinia megarhynchos*)[1****]

General Description
From the rear view, the nightingale is rather like a song thrush in both colour and shape; the tail is certainly more rust-hued but the back colour can be very similar. Considering the amount of romantic poetry which has been written about this bird it is quite a dull specimen; so many disappointed spectators have remarked on its sombreness as compared to its wonderful reputation for song. The upper parts are a rich russet shade, deeper rufous on the tail and darker brown on the wings. The male is white at the throat and more greyish of chest, fading again to white at the vent. The female is more of a creamy-pale brown on the chest, fading slightly at the throat, and white towards ventral area. The length is 17 cm (6¾ in).

Habitat
I suppose I was about 5 years old when I hand-reared my first nightingale; that bird later sang whenever his mealworm tin was placed in front of his flight. To this day I still have a fondness for this bird and its haunting song qualities when heard in the very early morning or breaking the stillness of night. The copses and spinneys and broad-leaved woodlands where overgrown bramble and nettles exist are its favourite haunts. The nest is usually placed on a thick platform upon the ground, in very low bramble or nettles. It is made of a dead-leaves foundation – oak are usual – and completed with more dead leaves, grasses and animal hair for the lining. The nightingale is a migrant.

Aviary
Although the standard minimum-sized aviary will suffice once a pair have accepted one another, this is another species which greatly benefits from the provision of two adjoining aviaries with a door in the dividing partition. Keep them next door to each other until you see the female showing her acceptance by remaining close to the division on the ground. When the male sings and displays she should respond. These birds will need a frost-free shelter attached for their use in winter. Heat is not necessary so long as extra lighting enables them to feed late into the evening and in the early morning.

Food
As for the black redstart (page 201), but include more fruit in the diet and Farex (Glaxo).

Breeding

The base of an aviary for nightingales' use should have a shallow pond, as they are frequent bathers. The rear half of the floor should hold two rhododendrons, or similar evergreens, beneath which should be a dense mass, at ground level, of honeysuckle, periwinkle and strawberry; these plants form just about the right density and are less unpleasant than brambles. A trip into some woods will normally provide a good-sized bag of old oak leaves; offer these, grasses and animal hair; cow hair or dog combings prove acceptable to the nightingales.

On the last occasion on which I bred these birds we had a crowd of friends gather at my home prior to a meeting in a nearby hall; the hen nightingale had young and I believe every one of the visitors peeped at her. They can become very tame at this time in an aviary and that particular hen took a number of insects from the proffering hands that afternoon.

When young hatch, increase the supply of cut-up mealworms and continue to do so gradually, all the time offering large quantities of other insects; they will accept small locusts, moth larvae and ant 'eggs' without any hesitation.

Thrush–Nightingale or Sprosser (*Luscinia luscinia*)[**]

General Description
This bird is darker brown than the ordinary nightingale and lacks its rufous shades. It also has a lightly spotted breast and is noticeably thicker-set; this stouter appearance gives the impression that it is far larger but the length is the same, at 17 cm (6¾ in). The song lacks the liquidity of the ordinary nightingale's. The young are even more heavily spotted than its near relative.

Habitat
This is much the same as that of the nightingale which is known in the British Isles, but, hailing from eastern Europe, the former USSR and Siberia, it has also a tendency to frequent more boggy, moist areas. They nest in the same manner with more or less similar materials.

Aviary
As for the nightingale (page 206).

Food
As for the black redstart (page 201), but include more fruit in the diet and Farex (Glaxo).

Breeding
I would treat as for the nightingale (page 207), although my experience is limited to one pair of these birds.

Siberian Rubythroat (*Luscinia calliope*)**

General Description
At first glance, when seen from the rear, this bird resembles the nightingale but, when seen from the front, the throat will immediately identify it. From the side, it will be seen to be shorter and less of a typical thrush shape. The length is 14 cm (5½ in) and the over-all colour is a dark wood-brown with an olive wash. The throat of the male during the breeding season is bright red and there is a white stripe above the eye from the forehead and a moustachial stripe of white; the lores are black and the scarlet of the throat is edged with deep slate-grey, fading to brownish grey, to buff, then to white at the vent. In the autumn the red throat of the male fades. The female has no red throat at any time and the eye and moustachial stripes hardly show at all; her throat is white. The juvenile bird resembles very much a young robin or redstart, being very spotted and an all-over brown.

Habitat
A migrant and a bird of wooded marshy areas, it tends to skulk in dense thickets and is found as far north as the Arctic Circle. It feeds mainly on the ground and nests in the shelter of a bush, upon the ground, or perhaps in densely growing plants. The nest is an untidy structure of grasses, a little moss, some rootlets and hair.

Aviary
Provide a 2 m^3 (6½ ft^3) structure with a frost-free shelter attached; the latter would need to have electric lighting during the winter months for extended feeding hours and possibly a little gentle warmth during the hardest of the weather. The outdoor aviary, into which the bird can be allowed daily, even in winter, should contain a small pond surrounded by a rather damp area, with moss and a few rocks. If this is in the foreground and the vegetation here is kept short, then the rear of the aviary could be allowed to grow fairly wild, with decorative grasses and

hops on the aviary sides. The section between the pond and the over-grown portion could contain a mixture of rhododendrons and azaleas for added seclusion and shelter. The part of the aviary joining the shelter should have a section of the roof covered to shield any nesting site from heavy and driving rain.

Food
As for the black redstart (page 201).

Breeding
The song of the male when he is courting the female is very sweet and melodious, though not as loud as a nightingale; it is rather soft, but somewhat unvaried. I have never seen any bickering between pairs of this species and have housed them together, but I have known other bird keepers to have trouble pairing them. For this reason, perhaps for the first attempt it would be prudent to pair them using two adjoining aviaries until their temperaments are known. Offer the nesting materials as recommended above and watch for some response from the female, usually a following of the male backwards and forwards along the dividing partition. My own birds got on remarkably well together and nesting took place almost as soon as I liberated them into a breeding enclosure from an indoor flight where I had wintered them. The eggs are a greenish-blue with pale reddish-brown spots on the larger end. For feeding during rearing treat as for the nightingale (page 207).

Red-flanked Bluetail (*Tarsiger cyanurus*)*

General Description
This is the size and shape of a redstart, measuring in length 14 cm (5½ in). Head, cheeks, mantle, back and tail are all bright blue, with lighter eye stripe, shoulder patch and rump. The throat is white and the breast is greyish, fading to white at the vent. The flanks show a rufous patch reaching half-way along the wing. A female is very nightingale-like in general coloration but she has the rufous flanks and blue on her upper tail. The juveniles are very like the young of robins, redstarts or nightingales, heavily speckled but with the upper tail blue.

Habitat
This migrant breeds mainly in the former USSR, frequenting the swamps of coniferous forests or boggy areas of birch wood. It is fond of low cover.

The nest is usually similarly placed to that of the robin, in bank cavities or a root of a fallen tree.

Aviary
My experience with this bird is rather limited and may not be typical but I have found it as troublesome as the redstart during pairing. To prevent a threatened loss I felt compelled to use two adjoining aviaries with a door built into the wire division; using this system one has control, which is always wise with so rare a subject. Plant the aviary as for the bluethroat (page 204), allowing a pond to overflow on to surrounding soft soil. I allowed my birds the use of outdoor quarters during the winter but provided them with extra feeding time by using a time-switch on the shelter lighting circuit; this shelter was not heated but seldom dropped much in temperature during the coldest of nights because of its good insulation.

Food
These birds enjoy taking mosquitoes from the pond in their enclosure; these are taken on the wing in a manner reminiscent of a flycatcher. Provide a diet as for the bluethroat (page 205), which is rich in animal protein, by adding ox heart, liver and possibly steamed fish and fish roe. Keep mealworms to a minimum, those that are given should always be cut into several pieces and well mixed into the food. Elderberries are taken when fully ripe.

Breeding
Provide the biscuit-tin-type nest site, securely fixed on its side to a low pole and filled loosely with bracken. As an alternative, offer a half-open-fronted nest box. It is all very well to build natural sites into a decorative bank or rockery but, with the difficulty of keeping naturally planted enclosures totally free of mice, it never seemed to be worth risking with such a rare bird. The male will attempt to attract the female to a site in much the same way as the robin and, although the male will carry nesting material to the site, it would appear that the nest is made entirely by the female, using moss, grass, rootlets, animal hair and wool.

In the event of any young being obtained, feed as for the black redstart (page 201).

Robin (*Erithacus rubecula*)*****‡

General Description
For anyone who may be unfamiliar with this bird, it measures 14 cm (5½ in) in length. Its distinctive orange-red face and breast, white belly and vent, olive-brown upper parts, erect stout-looking stance and its confiding manner in visiting gardens during the winter should help even the complete stranger to European birds to identify it. It can be very pugnacious when guarding its territory and will not even tolerate a member of the opposite sex during out-of-breeding-season times. Two males will fight to the death over a mutually chosen territory and I have personally separated two by hand, when the uppermost was pecking at the head of the trespassing bird, and I was compelled to keep the attacked specimen until the head healed over and I could release him in the country.

Habitat
Woodland, park, garden or hedgerow, the robin is common in most places, nesting in a bank cavity, a shelf in a garage where free entry can be obtained at all times, the root of an upturned tree; I have even known them to nest in an empty open aviary.

Aviary
Due to their pugnacity even towards members of their opposite sex, this is another bird which I would recommend be housed in adjoining aviaries prior to peak condition being reached. Each need be the standard minimum size only. There is no need for anything but the open-fronted type of shelter joining the aviary at one end as these birds will winter out quite well. In fact they will come into condition earlier than if housed indoors during the colder months. There should be a perch at each end of the enclosure, with two more in the shelter, situated fairly high. The food tray should be in an easily accessible position so that later, when they are breeding, one does not need to enter the aviary. A small pool will be welcome since these birds will bathe even when snow is on the ground. Apart from this, it will encourage insects to visit their enclosures.

Food
As for the black redstart (page 201).

Breeding
Place a nest box, of the half-open-fronted type, in each aviary with each

sited as near to the dividing partition as is possible; if well roofed and covered it will not matter if they are in the open: I would suggest they are placed at a height of 1 m (3¼ ft) from the ground. The male, upon reaching breeding condition, will sing constantly and display to the female; until she responds they must not be placed together. When she reaches the required condition and she sees the male fly to the nest box in his enclosure and enter it, she will then try to follow him by clinging to the wire division as near to him as she is able. The second day they will do this again and one or both may carry a dead leaf or some other piece of nesting material; when this occurs, open the dividing door and keep a watchful eye on their general behaviour. All should be well from this point onwards, even more so if a wild pair are in the vicinity. The male under controlled conditions will then challenge the wildling and quite possibly have a daily fight through the wire with him. This is a very natural event and would be happening in the wild; it deflects much of his aggression and channels it in the right direction, towards any interloper.

I would think that the robin is one of the simplest of insectivorous birds to breed, in that where the provision of suitable rearing food is concerned they make devoted parents. Once the young become self-supporting, however, make every haste to remove the young birds as, at that time, it is quite natural in the wild for the male to drive them out of the territory so that his mate can nest and rear another brood. Many birds among the softbill species have this tendency, and it will be found necessary to carry out this procedure in the majority of breedings. Feed as for the nightingale (page 207) when the young hatch.

Siberian Blue Robin (*Luscinia cyane*)

General Description
A very rare bird to find in aviculture today, its upper parts are dark blue, rather matt in appearance, although the forehead is far brighter; the tail, wings and crown are brown but washed with blue, the underparts are white and the flanks are rufous-flushed. The female is more olive-brown, with a slight rufous tint showing in the wings and tail and the underparts are white. The length of this bird is 14 cm (5½ in).

Habitat
It is found in Siberia, among dense bushes and undergrowth where it creeps and skulks. It sings well, from a low tree or bush, nesting upon the ground in dense vegetation.

Aviary
As for the rubythroat (page 208).

Food
As for the black redstart (page 201).

Breeding
The only two birds I ever had caused considerable worry if placed together during the winter months. To avoid disappointment with such a rarity, I would advise pairing them by using adjoining aviaries as described for the robin (page 211).

Wheatear (*Oenanthe oenanthe*)[1***]

General Description
In breeding plumage the male can be a strikingly attractive bird with pale powder-grey crown and mantle, black cheeks beneath a white eye stripe, white throat, and an upper breast of a warm buff fading to white at the vent; the wings and tail are black and the rump and tail coverts are white. The female is much browner and paler, more of a sandy-buff, but retaining the white rump; in winter the male resembles the female. The length is 14.5 cm (5¾ in).

Habitat
The wheatear inhabits open country, uncultivated areas and moorland.

Aviary
With the knowledge that the pairing of these birds can sometimes cause problems, one should be prepared for such eventualities; it is far better to use two enclosures side by side with an adjoining door, than place two birds together and hope for the best results to occur. I would suggest as a minimum for the breeding aviary a 2 m^3 (6½ ft^3) enclosure; the adjoining one, for the use of the female until breeding condition and pair acceptance are reached, need be only half this size. Since these birds will use rock crevices, rabbit holes, drainage pipes, cavities in heaps of rubble and the like, it is not taxing to fabricate a site to suit their needs. Make certain that the site is not on ground which is likely to become waterlogged or which is in the path of heavy downpours of rain. Construct a main nest chamber of bricks or fairly thick timber, without a top and one side. Run to this a piece of earthenware pipe, approximately 12–15 cm (4¾–6 in) in diameter and about 30 cm (12 in)

long. Cement all this in position then cover it with soil, except for the roof of the nest chamber; for this one should construct a lid with a rim, like a biscuit-tin lid; make it of timber and cover it with roofing felt and make sure that one can raise it easily for inspection at a later date. Wheatears will benefit from having an enclosure to which they can retire during winter months, though they may be allowed the freedom of the outdoor enclosure. Make sure their aviary floor is well drained; the use of plenty of sand, shingle and gravel during the preparation stage will ensure a maximum of dryness. With the addition of heathers, rock plants and a song post at each end all will be ready for the occupants.

Food
A good insectivorous mixture with grated cheese, steamed ox heart and the usual minced vegetable matter will form the basis. Give very little fruit other than in autumn when elderberries will be taken. Include a small daily quota of cut-up mealworms in the food; items such as wax-moth larvae and particularly locusts, of which they are extremely fond, can be given daily in small quantities; these should be increased when the rearing of young is expected.

Breeding
As the male advances in condition so he will come into song and try to call the female to one of his selected sites. When, and only when, she is observed making a response and trying to reach the nest site, open the dividing door. When paired in this manner things should progress satisfactorily. Many birds, both male and female, are lost through being introduced before they are in peak condition. The materials they will require for nesting are grasses, moss, rootlets, dead leaves, small tufts of wool and animal hair; dog brushings are ideal.

With the hatching of eggs, increase the number of chopped-up mealworms in the food; double will suffice at the first stage. Offer ant larvae if obtainable, newly hatched locusts, wax-moth larvae and greenfly; almost any small garden insect, woodlice and earwigs will all find a ready welcome.

Greenland Wheatear (*Oenanthe oenanthe leucorrhoa*)[***]

General Description
I have measured these birds at times and found them well over the expected length, at 16.5 cm (6¼ in): this is considerably more than the wheatear, the more common bird. In the field they can look very similar,

but close by the Greenland wheatear is found to be far more bulky of build. It is altogether a darker bird with a darker and richer grey on the mantle; the wings too are blacker, and the breast is almost tan compared with the buff of the smaller race. The plumage is far paler during the out-of-season months, but even at this time differs considerably from the common bird.

Habitat
Although called the Greenland wheatear, this bird nests over a vast area of Greenland, Iceland and north-east Canada, tundra areas and open wastes being its haunts.

Aviary
As for the wheatear (page 213).

Food
As for the wheatear (page 214).

Breeding
As for the wheatear (page 214).

Black Wheatear or Black Chat (*Oenanthe leucura*)

General Description
One of the larger birds of this family, the male in adult plumage is mainly black, with white just on the rump, outer feathers of the tail and beneath the tail coverts. The female is much browner. They are at times referred to as black chats and are very striking birds to have in an aviary. The length is 18 cm (7 in).

Habitat
Typically, this wheatear belongs to the western Mediterranean and frequents rugged rocky areas, cliffs on coasts, quarries and even old ruins of buildings.

Aviary
The enclosure should be no less than 2 m^2 (6½ ft^2) and of a similar height because the bird is very active and its behaviour demands roomy quarters, particularly for courtship and breeding. Certainly its full beauty cannot be appreciated under more confined conditions. One of our pairs lived in south Buckinghamshire throughout one winter, although they

did have access to a feeding station beneath a roof and were exceedingly well sheltered. In the immediate vicinity of this food supply was a small-wattage electric light, quite adequate to illuminate the area well in darkness, and they were often seen eating and roosting nearby. A further pair spent the winter in a bird-room flight, yet with the arrival of spring the birds which had spent the winter outside looked the better. Even so, over such a small area as the British Isles, the weather conditions vary alarmingly, so I can hardly recommend keeping them outside as a general practice.

Food
My specimens, perhaps not numerous enough to form conclusions about food mixtures over many years, were very fond of hard-skinned insects, which provided much shell etc. for roughage; items such as mealworm beetles, mealworms and earwigs were never ignored. During the summer months they took anything that moved – wax-moth larvae, green aphids and spiders, and were even observed wading in shallow water after tadpoles. Softfood contained a slightly higher proportion of animal protein to vegetable matter but we still included grated apple because, although they invariably left some of the chopped greenfood, everything else disappeared. Part of their aviary covered one end of the garden goldfish pond and, despite their best attempts to ruin the frogs' breeding programme, many of the minute frogs escaped to the main section outside the wheatear enclosure.

Breeding
Since these birds, when nesting in the wild, use rocky recesses, gaps between boulders or even fissures in rock faces as sites, a half-open-fronted site made of wood will be accepted, as will a number of well-hidden sites positioned behind one small entrance. I have watched two hens, in separate enclosures, carrying nest material to two man-made box-type sites, well over 1 m (3¼ ft) from the aviary floor. A length of glazed earthenware sewage pipe, with ends well hidden, could easily appeal to them, or rocks, which form an almost natural site – beware of nocturnal visits by mice. Watching this species feeding in the wild, I even saw very small lizards taken; these were hammered upon rocks and were soon reduced to limp, soft edible morsels. Beetles, grasshoppers, locusts etc. were all taken whilst under observation. These wheatears readily accept man-made foods as supplied commercially and the same goes for acceptance of the live insects offered. To a fully experienced keeper of insectivores, the breeding of this bird should therefore cause no greater problems than the common wheatear.

Thrushes, 'Chats', Redstarts, etc.

Black-eared Wheatear (*Oenanthe hispanica*)

General Description
When in breeding plumage the male is very attractively coloured; the wings and tail are typical of this family, black but with the usual white coverts and rump. The ear coverts are also black; the black commences at the bill and a black line runs through the eye in much the same manner as the cheek patch of the common race but here it extends further back. The body colour is a creamy-buff and the crown at times is almost creamy-white. The female is pale brown, almost like the common female. There is also an eastern race which has a black throat. The length is the same as the common race at 14.5 cm (5¾ in).

Habitat
Barren dry open country is the favourite terrain, as well as vineyards, particularly those with stone walls.

Aviary
As for the wheatear (page 213).

Food
As for the wheatear (page 214).

Breeding
If a wall can be constructed with natural-looking holes, perhaps formed by a terracotta flowerpot cemented into position with a jutting slate over-hang for protection from rain, these birds can be persuaded to use them. Although these birds tend to nest at ground level, even beneath dense shrub or vegetation, many will still take to hole sites and thus keep better out of the way of mice. For all other details consult the wheatear entry (page 214).

Alpine Accentor (*Prunella collaris*)****

General Description
This attractive specimen is far larger than the hedge accentor and more brightly coloured and, although predominantly brown, grey and chestnut, its markings are beautifully uniform in appearance. The throat is dappled white and black and the flanks are chestnut streaked in a rich rufous shade. The song is sweet-sounding and constant, a continuous musical warbling, the first few notes reminiscent of a piece from the

woodlark's song. The sexes are very much alike; in fact I have mistaken one for the other quite often. The juvenile shows far less of the rich colours and is more greyish, having no throat markings or rufous on flanks. The length of this bird is 18 cm (7 in).

Habitat
A bird of the high European mountains, when seen, and this is only rarely, it is usually on some rocky slope among juniper bushes and stunted conifers where other vegetation grows thinly. It nests in a low bush or rock crevice beneath such growth.

Aviary
I would like to think that a minimum of 3 m (9¾ ft) long by at least half that in width and 2 m (6½ ft) high would be given as an enclosure for a pair of these birds. They can live together all the year round and need no shelter other than a small-roofed area at each end, this dropping downwards around the sides for more complete weather-proofing. They seem to greatly appreciate rocks, so the base can be decorated in many ways using this material and also suitable plants; a juniper or two and conifers will give them a perfect setting. Bathing facilities, such as a shallow pool, will be made full use of.

Food
The food should include both seeds and softfood together with some insects. I use threshings plus rape, panicum millet, maw and gold-of-pleasure; they enjoy the grasses as much as anything. An insectivorous mixture with a little grated cheese and a few cut-up mealworms should be given daily; often the consumption of this during the winter months drops considerably but it increases again with the arrival of spring, summer and autumn, before lessening again. Mealworms will be eaten during the winter period.

Breeding
Against all expectations and despite specially prepared rock crevices and nest boxes, our birds nested low in a dense buddleia bush and reared successfully there. They do not seem very territorially minded and, when I had two pairs together at one time, there was little bickering.

The nest looks rather like a dunnock's but is larger and made of grasses, a little moss and lined with animal hair. It was sited about 30 cm (12 in) above the ground. The eggs are light blue but with a matt greyish wash.

If they gave any seed to their young I never witnessed it, but they did regurgitate when the young were very small. They certainly collected aphids. The young were out of the nest well over a week before they were seen picking up food, which in this case quite likely included seeds, but I could not be sure. They resemble the hedge accentor in feeding habits and make one wonder at the actual food intake. We gave a wide range of livefood when the young were in the nest; wax-moth larvae, mealworms, caterpillars, in fact anything found crawling, was usually offered. There was a continuously running stream in their enclosure, and young and old were often to be seen at the edge, eating something either from the moss there or from the water itself.

Black-throated Accentor (*Prunella atrogularis*)[†*****‡]

General Description
This is a rather sombrely clad bird of black, brown and a light shade of mushroom; the head is black with a clearly defined light brown eye stripe and moustachial stripe. The throat-marking ends abruptly on the chest which is of light mushroom. After the breeding season the black tends to fade quite a lot. Very typical of the accentor in behaviour and movements, it has a quick, flitting, flight and is always on the move in undergrowth and at ground level. The length is 14 cm (5½ in).

Habitat
Another bird of the high mountainous regions, mine came from the former USSR, where stunted conifers and juniper grow. Nesting takes place almost at ground level in a dense bush.

Aviary
The standard recommended minimum enclosure should be sufficient for these birds. It can be planted out, rockery-style, with junipers and conifers around the edges. A small shallow pond is used daily when available.

Food
As for the alpine accentor (page 218).

Breeding
The nest, which was discovered after the young bird had fledged, was found deeply set in a cluster of dead bracken, clumps of which had been fastened at intervals almost from ground to roof on one side of the aviary. It was about 30 cm (12 in) from the ground and made of dry grass and

chickweed, then lined with animal hair. Two broken egg shells were found, in colour seeming very like those of a hedge accentor but perhaps more glossy. In all respects treat as for the alpine accentor (page 218).

Hedge Accentor or Dunnock (*Prunella modularis*) ****

General Description
Of a skulking nature, with a quite attractive song, this bird is always tight of feather when all else in the depth of winter looks forlorn. Active and lively, it is a brown bird with a greyish head. The male has grey underparts and an even more slate-grey head during the breeding season. The length is 14.5 cm (5¾ in).

Habitat
It is found almost anywhere in gardens, woodland, country lane, vineyard, farm, even town park and any open space cultivated or wild, throughout Europe.

Aviary
The standard minimum is sufficient, planted with bushes at the rear. Include a shallow pond for bathing and to attract livefood. The merest of roofed sections will be sufficient against rainy days or winter shelter.

Food
As for the alpine accentor (page 218).

Breeding
Probably one of the simplest of native birds to breed in captivity, although common, it makes a pleasant bird to study. Often the male will complete a further nest whilst his mate is still feeding the first lot of young. I have known a male build five nests in one season in this manner and the female laid in each as her young of the previous nest fledged, the male taking over parental duties in respect of the fledglings. One bad point with accentors is that the male will resent the presence of the young and continually chase them; they should therefore be removed early so that fatalities do not occur. These birds will nest in almost any small bush, hedge, shrub or climbing honeysuckle. The nest is made of twigs, grass, leaves and moss with a lining of animal hair. The eggs are glossy bright blue.

11

Larks, Pipits and Wagtails

HERE WE HAVE A GROUP of fairly coarse feeders, all spending much of their time upon the ground where they search for insect life. Larks will take a number of small seeds during the winter months. Although aphids and small insects will be greedily taken whenever the opportunity arises, these birds are well able to take far larger insects and I have seen them all tackle almost fully grown locusts: the young of these insects are certainly acceptable to this group for feeding their young. Many wagtails in captivity have been reared, for example, on small fish fry; I mention this so that the reader will experiment and investigate further possibilities. The fish mentioned were placed in shallow containers holding a small amount of water.

The basic food should be a nourishing man-prepared insectivorous mixture, with the additives mentioned in Part One on feeding. The mealworms can be increased slightly in number but should still be cut into small pieces and mixed with the food.

With this group of birds, due to their habits of ground feeding and running around hunting, one has a wonderful excuse to make the aviary into a real show-piece; small ponds, rockery plants, decorative stone and rocks, clear areas of flat shingle and sand, mossy banks to the pond and flowering shrubs will help in bringing to life what is far too often a mere dead area of wire and soil. Some of the aviaries I have seen holding the wagtail family have included waterfalls, running streams, much semi-tropical vegetation and even fish pools; apart from the visual pleasure one can obtain from such garden surroundings, the natural insect life is an added bonus for the inmates.

Providing that larks have been housed together all the winter, they can be introduced to a breeding pen together. With pipits, a certain amount of rather hectic chasing is quite normal so do not be too distressed to see a male singing violently and chasing a female back and forth in their enclosure; they too can be wintered together and introduced to the outdoor breeding aviary at the same time.

Shore Lark (*Eremophila alpestris*) ***

General Description
One of the most attractive larks, it measures 16.5 cm (6½ in) long and is distinguishable from all others by the yellow fore-throat and patch to the rear of the black cheeks. It also has distinctive black 'horns' rising up from each side of the head, above the yellow of the forehead which recedes over the eye and down behind the cheek patch to link up with the throat. There is a black crescent on the upper chest. The over-all appearance elsewhere is of warm brown with perhaps a slight pinkish wash, the upper parts being more noticeably so; the underparts are much paler. The female shows less black and, upon close inspection, will be found a little smaller.

Habitat
It is a bird of the Arctic Circle where it breeds in the mountainous alpine regions, wintering south on salt marshes and beaches.

Aviary
I have kept many of these birds and each one has become tame to the point of foolishness, walking over and around my feet in their enclosure; yet despite this fearless and tame attitude in captivity, the mating display flight does require a larger area than may first be considered necessary. Make an effort to provide at least a 3 m (9¾ ft) length by 2 m (6½ ft) width and height. Such an enclosure can be made most attractive and provide ample space for their mating and breeding needs. A small roofed-in section at each end, with one or two downward fixed boards, will keep them dry in the wettest of weather and provide a reasonably draught-proof roost. There need be no special winter shelter but, if one can be attached, then I feel sure it must contribute to longevity. Our birds nested beneath a tuft of coarse grass growing under, and up through, a small conifer, so keep the aviary base a mixture of peat and shingle with heathers and a few small, stunted or horizontally growing conifers.

Food
Provide a good softbill mixture, with the usual additions of grated cheese, vegetable and a few cut-up mealworms. Although it is basically a softbill, screenings or small wild seeds can be strewn over a small section of the aviary floor; grass seed and the smaller of the brassicas will be pecked over during the winter months but these should be provided only as a supplement to the insectivorous food. Because of their taste for small

crustaceans, I often minced up a few shrimps or prawns in their food and on such occasions it was noticeable that they were cleared completely by the next feeding time.

Breeding
In the wild this bird nests beneath a jutting rock, in a cavity in the ground or in some similar site that is usually shielded by a tuft of vegetation; I have never had one build other than on the ground. The materials were merely a variety of dry grasses, bents and a lining of vegetable down, and the structure was rather untidy. My birds took fairly large quantities of young locusts, although they were presumably foreign to them as an item of food, and wood-ant larvae; fly maggots were given after being thoroughly cleaned and treated with cod-liver oil and Vionate (Squibbs). Wax-moth larvae, enclosed in the material which they use for pupation, were given by the handful, and the birds would tear the silk-like material into shreds in order to reach the insects inside.

Short-toed Lark (*Calandrella cinerea*) ***

General Description
Measuring 14 cm (5½ in) in length, this lark is a sandy-brown in colour, lightly streaked with darker brown and with white underparts; it displays a small dark patch on each side of the breast. The sexes are very alike but the female is slightly smaller. Neither can be called stocky or thick-set but the female is certainly the slimmer of the two.

Habitat
It inhabits sandy wastes, dune-like areas and steppe country in southern Europe.

Aviary
The minimum standard enclosure will house one pair comfortably. Keep the base well drained with a finish of fine shingle and sand. A few heathers and tufts of sedge or long grass will provide them with an ideal setting. A wintering shed should either be attached or available for their habitation during the colder months. Extra feeding time will greatly benefit them during such times.

Food
Feed as for the shore lark (page 222).

Breeding
Although the nest is apparently started in a hollow of the ground, it ends up looking as if it were on a platform of smaller stones; this is an illusion since the stones are piled up to the edge after the structure has been completed. At the foot of the clumps of grass, make impressions in the shingle with the clenched fist, then hang the vegetation over to hide it. The materials this bird will need are mainly dry grasses but there is a final soft lining made with animal hair, vegetable down and feathers. Any resulting young should be offered insects as for the shore lark (page 223).

Lesser Short-toed Lark (*Calandrella rufescens*)

General Description
This resembles the short-toed lark to a great extent but is far greyer about the upper parts and also darkly streaked on the upper breast and without the dark chest patches. It is not crested in the true sense but does tend to raise its head feathers. The length is the same at 14 cm (5½ in). The sexes are alike but here again, when one can compare a true pair, it will be seen that the female is slightly smaller.

Habitat
This is very similar to that of the short-toed lark but it seems to show a preference for damper vegetation and marshy areas adjoining water.

Aviary
As for the short-toed lark (page 223).

Food
As for the shore lark (page 222).

Breeding
As for the short-toed lark (above).

Wood Lark (*Lullula arborea*)[†*****]

General Description
This bird is shorter than the skylark and lacks the white outer feathers in the tail; otherwise it is rather similar. The plumage is a rich brown and it has a whitish-buff eye-stripe running to the nape. The breast fades

to white on the underparts and the head feathers are often raised in crest-like formation but not always so. The sexes are marked alike, the darker streaked back of the female being a little less broad in comparison. The length is 15 cm (6 in).

Habitat
I prefer the song of this bird to most others I know and, when listening to it in wild open country, quiet orchard or on alpine summer farmland, it has a wonderful quality hard to describe. One never quite knows how one may come across this bird, whether singly, in pairs or as a family group, in which they travel very often. They may be seen from the New Forest in the British Isles to the olive groves of Italy.

Aviary
Give one pair an enclosure of at least 3 m (9¾ ft) long by 2 m (6½ ft) in width and height. Erect a song post for the male where he may sing in the sun; this need be only a rustic pole standing 1 m (3¼ ft) above the ground. The base of the aviary should be kept fairly clear; allow two corners to become a little overgrown with grass, but elsewhere it is better to have a hard shingle or gravel base, with perhaps a few heathers. As a wind-break and to attract insects, plant hops at the rear of the aviary. This bird may be a resident in the British Isles in the winter, and I have seen them nest whilst there is snow on the ground, but a frost-proof shelter with extra light to extend the feeding hours should prevent any losses and enhance one's chances of breeding these larks.

Food
As for the shore lark (page 222).

Breeding
As for the shore lark (page 223). Nesting materials should comprise moss, coarse and fine grasses and animal hair.

Skylark (*Alauda arvensis*) ****

General Description
In length the skylark is 18 cm (7 in). The mantle is a duller brown than the woodlark and there is a far longer tail, with white outer feathers. Here again the crest is not always raised but can be very conspicuous at times. Head and mantle bear dark streaks, the underparts are whitish with streaks on the upper breast and the eye stripe is a little more buff

than in the woodlark when compared side by side. The sexes are alike except for the male's heavier build.

Habitat
Open country, cultivated or fallow, provides haunts for this bird; it may be alpine meadow or low heath or fen. On a warm summer's day one may be lucky to hear this bird's song somewhere in the vicinity.

Aviary
As for the shore lark (page 222).

Food
As for the shore lark (page 222).

Breeding
As for the shore lark (page 223).

Meadow Pipit (*Anthus pratensis*)***

General Description
Although dark brown, the meadow pipit tends to appear greyish-brown when compared with the reddish-brown of the tree pipit. It is streaked with darker brown on the head, mantle and upper chest, the underparts are whiter and the back claw is long and almost straight, far longer than that of the tree pipit, which curves more for perching. The tail has white outer feathers which show very clearly in flight. The length of this bird is 14.5 cm (5¾ in).

Habitat
It frequents uncultivated country, dry or damp, sand or bog, and is always on the move in low pasture land. It is found along freshwater margins and is not dependent upon trees.

Aviary
A pair should have an enclosure of at least 2 m (6½ ft) in length; the mating flight can be boisterous so if one can manage to build the enclosure 1 m (3¼ ft) longer it would be more suitable. Allow 1.5 m (4⅞ ft) width and 2 m (6½ ft) in height. Include a shallow pond in their aviary and the remainder of the base can be grass, except for a small perimeter around the pool which should be shingle. Winter quarters should be attached, with lighting to facilitate extended feeding.

Food
A basic insectivorous food, with added grated cheese, minced vegetable and root crop should be provided. A few mealworms cut into pieces and mixed well into the food will be appreciated. Give daily feeds of any available live insects but not in any great numbers unless with young.

Breeding
The nest will be made from various grasses and vegetable matter and then lined with animal hair. When young are being reared offer additional mealworms, cut up in the food, ants' larvae, moth larvae and young locusts. Fly maggots can be used after treating as usual with a little oil and additives. The nests are sometimes so well hidden in the grass, just under an over-hanging tuft, that it can be exceedingly dangerous to enter the aviary when breeding is in progress.

Red-throated Pipit (*Anthus cervinus*) ****

General Description
Resembling the tree pipit more than any other of the family group, but with an attractive reddish throat when in full breeding plumage, this pipit has a far better song than either the tree pipit or meadow pipit and is rather musical in some respects. This plumage is a rich brown and the bird's energetic little sprints upon the aviary floor are very similar to the running motion of the more common pied wagtail. The length is 13.5 cm (5¼ in).

Habitat
It is found in northern Europe and Asia, where it occupies a terrain very much similar to that chosen by native British pipits.

Aviary
Our specimens were hand-reared in northern Europe and came here as adult birds. They had never been in an aviary before, having lived in a large cage in a bird room until we received them. We provided an enclosure for each pair that measured a little less than 3 m^2 (9¾ ft^2). Because of their upbringing and the manner in which they had been kept, they were remarkably tame, actually walking over our feet in the enclosure as we fed them with mealworms, and thinking nothing of taking food from our fingers. I feel that, had we given these birds a 2 m (6½ ft) by 1 m (3¼ ft) enclosure as a minimum, similar results would have been produced and the behavioural patterns would have been the

same.

Food
Once again, this pipit resembles the British ones closely in its choice of foods, taking readily to our usual softbill mixtures with appropriate additives; their management, therefore, need not be varied at all. In addition to softfood, we gave a dish of seed consisting of pinhead oatmeal, ground peanuts, shelled millet, as used for human consumption, and a foreign finch conditioning seed. During our observations, they always seemed to select seeds minus a husk and, in particular, the lighter-coloured varieties, but softfood still outweighed seed intake and formed the bulk of their food.

Breeding
It was in 1961 that one of our pairs nested. The other pair, despite being within both sight and sound, a mere 3 m (9¾ ft) away, which is nothing to a bird, did little other than carry nesting materials. Perhaps this was due to the male of the nesting pair having a far more dominant attitude at all times, even when the four wintered altogether in one flight; his near presence may well have intimidated the milder natured specimen. Fine grasses of a very dry nature formed the bulk of the nest, some rootlets were included and the interior was lined with dog hair; nothing else was found later when the nest was dissected, although we provided feathers, wool and plenty of vegetable down. Four eggs were seen in the nest. For want of a better way to describe them, I would say they were of a mushroom-olive hue. The closest I approached their nest, for fear of offending the breeding birds, was about 1 m (3¼ ft) and, even then, I used a thin bamboo cane to raise the grass and coltsfoot under which they had built it, in order to check on the contents. Even though many large comfrey leaves were hanging, and seemed a suitable site to me, they had accepted a far smaller leaf – the coltsfoot – for their nest site. My observation of their nest and general behaviour was rather casual and distant. I noted that incubation was about 14 days but could not swear to an exact period; I may well have been a day either way in my calculations. Only two chicks were reared to maturity; although four hatched, the other two disappeared much earlier.

Rock and Water Pipits (*Anthus spinoletta*)*

General Description
Oddly the Latin name is shared in this instance. The water pipit is the paler, rather more grey, and has white outer tail feathers. In these two respects it differs from the rock pipit, which has grey outer tail feathers and is a little more brownish, though less so than either the tree or meadow pipits. Of typical pipit appearance, these races can at first be confused one for the other but the more one handles them, the greater seems the difference, though the textbooks usually differentiate between them merely on habitat. The bird that frequents the mountainous areas is referred to as the water pipit and the one inhabiting rocky coastal regions is the rock pipit. They are both streaked on the upper breast and mantle, the belly being whitish-buff and the water pipit having a distinctly noticeable white eye stripe. These races are the largest of our European pipits, measuring 16.5 cm (6½ in) in length. The sexes are very much alike.

Habitat
The rock pipit is more or less confined to rocky stretches of coast, only coming inland in winter during the worst of the weather. The water pipit inhabits the mountainous areas of Europe and tends to have the more southerly range.

Aviary
Both races use a rather hectic mating flight whilst chasing the female and for this reason an aviary a minimum of 3 m (9¾ ft) in length should be planned, with a width and height of 2 m (6½ ft). A depression in the ground, hidden by vegetation, is the normal nest site chosen, or perhaps a crevice, but I have known these birds to nest in a half-open-fronted nest box at varying heights from the ground. The ground should be shingled with a few decorative rocks and plants surrounding a shallow aviary pool, I have watched many times small freshwater shrimps being taken by these birds whilst wading, long-legged, in such pools.

Food
As for the meadow pipit (page 227).

Breeding
Offer coarse and fine grasses, moss, animal hair or fibrous string cut into short lengths and teased-out into single strands. In other respects, treat as the meadow pipit (page 227).

Tree Pipit (*Anthus trivialis*)[†*****]

General Description
A rich dark brown, this bird is darkly streaked on the head, mantle and upper breast; the background on the breast is a buffish-tan which fades to a white ventral area and the throat is also whitish. Sexes are very similar but it is normal for the breast streaks in the female to extend further down the belly and for the background colour of the upper breast to be paler. The hind claws of this bird are shorter than those of the meadow pipit and curved, making them more suited to its arboreal habits. The length of this bird is 15 cm (6 in) and it is stoutly built.

Habitat
This migrant is found on woodland heaths and commons with scattered trees. This is the arboreal member of the pipit family and it will have a favourite song point from which it begins its song; it will then rise steeply and quietly, bursting forth into full song as it reaches the summit of the song-flight before plummeting earthwards, singing the whole time, to land back on the same perch.

Aviary
Recalling the flight-song habits, and the mating chase which precedes true pairing, a 3 m (9¾ ft) long enclosure should be allowed; this can be 2 m (6½ ft) wide and of the same height. This type of enclosure will take a stunted silver birch at each end: if the tap-root is cut three times over its first 4 or 5 years and it is wired in position it should remain well within the limits of the roof. A small shallow pool, with a sand and shingle bank, and grass for the remainder of the base will take care of these birds' aviary requirements for the better part of the year. For the winter they should be given the use of a frost-free enclosure where they can venture outdoors as they wish. This shelter should be provided with artificial light, so enabling extra feeding time to be taken. In this manner and under such conditions tree pipits will live for many years.

Food
As for the meadow pipit (page 227).

Breeding
Supply with moss, dry grasses, bents and withered fibrous leaves such as iris, with finer grasses and animal hair for lining materials. In other respects treat as for the meadow pipit (page 227).

Grey Wagtail (*Motacilla cinerea*)†***

General Description
With blue-grey upper parts, the male has a black throat during the breeding season. Occasionally I have known females to have this throat marking as well. The underparts are a brilliant yellow and the very dark wings are tipped with white. The tail also has white outer feathers and both above and below the eye is a white stripe. The female is a little duller. The length of this bird is 18 cm (7 in). The juvenile birds are duller and slightly brownish on the chest, showing bright yellow only on the under-tail coverts; the young males are brighter than the young females and paler on the throat.

Habitat
Rarely does this bird wander far from water; streams and old country bridges, and old rock walls near streams, are all favoured. In summer it frequents fast-running mountain streams and hilly rock-bound streams; during the winter it moves to lower freshwater areas, even village parks where water exists. It nests in a wall cavity, where a brick may have fallen out, beneath a bridge, in a jutting waste-water pipe, or along a steep bank in the roots of a tree overhanging the stream.

Aviary
An enclosure should be about 3 m (9¾ ft) long by 1 or 2 m (3¼ or 6½ ft) wide and have an adjoining winter shelter where electric light may provide a little extra feeding time during the winter months. A shallow pond should be incorporated into this aviary, and, very near this, two or three near-natural sites be built for their use. These can be constructed quite easily with water-proof inspection lids: make certain that the interior is dry and not likely to be flooded by heavy rain. The inclusion of a few ornamental rocks, an evergreen, a few small flowering shrubs and plenty of moss planted around the edges of the pond should give a pleasant setting. These birds can be allowed to enter and leave the shelter all the year through.

Food
The basis will be the insectivorous mixture, and additions will be grated cheese, minced steamed ox heart and liver, fish and fish roe, cooked and minced in the same way, well mixed with the food. A few mealworms daily can be cut into small pieces and stirred into the mixture. The same treatment should be given to small locusts and moth larvae and any other insects given when no breeding is taking place. Supply daily small

quantities of greenfood, finely minced, but only occasionally does fruit need to be mixed with the food; they will, however, eat ripe elderberries during the autumn.

Breeding
These are fairly simple birds to breed from; and, in fact, most of the wagtails are ideally suited to somebody just embarking upon the keeping of insectivorous species. The nest materials to offer are fine twigs, rootlets, grasses, moss and animal hair; dog combings are perfect as they make a good nest lining.

With the arrival of young, double up on the number of cut-up mealworms on the first day they hatch; treble it on the third day, always cutting them into pieces and mixing them into the food. At the same time as this increase in mealworms, supply large quantities of small locusts and moth or wood-ant larvae. Feeding now becomes necessary three or four times per day; as the insects are used, so the containers need replenishing. Maggots should not be used before the young birds are about 5 or 6 days of age; then be sure they have been cleaned thoroughly and supplemented with the recommended vitamin/protein/mineral additives.

Pied Wagtail (*Motacilla alba yarrelli*)****

General Description
The most common of the wagtails, the contrasting black and white makes it a most attractive bird to watch. The crown, breast, mantle, back and tail are all black, and the more shining-jet-coloured it is, the better, normally, is the health of the specimen; the cheeks, forehead and underparts are white. The length is 18 cm (7 in), and it is rather more stout than the white wagtail, which it resembles so closely; it can be distinguished by its rump, which is black rather than the grey of the white variety.

Habitat
It is never very far from water, but although often frequenting rivers and fast-flowing streams, it is seen as often on cultivated farmland near a pond or boggy area. During the winter it comes into gardens and nearer human activity. It nests almost anywhere, from a garden-shed shelf, drain pipe, cavity or hole in bank or rock face, to an upturned root of a tree or a thatched cow barn, in the thatch or inside.

Aviary
As for the grey wagtail (page 231), but offer half-open-fronted nest boxes at varying heights.

Food
As for the grey wagtail (page 231).

Breeding
As for the grey wagtail (page 232).

White Wagtail (*Motacilla alba*)****

General Description
Apart from the grey rump of this bird and the mantle, which is of the same colour, there are few other striking differences between the white race and the pied during the breeding season. The grey back is the surest sign, whether in the field or aviary, but even so there have been many cases of suspected hybridisation so that some very doubtful cases exist. The pied should have a definite area of black, mantle to rump, during his breeding season.

Habitat
As for the pied wagtail (page 232).

Aviary
As for the pied wagtail (above).

Food
As for the grey wagtail (page 231).

Breeding
As for the grey wagtail (page 232).

Yellow Wagtail (*Motacilla flava*)†****

General Description
There are many variants of this bird, the head plumage differing very widely in some cases while in others handling seems the only way for true identification. It is a little slimmer than the other wagtails. When describing the yellow species, we have in mind the green-headed bird

but there are also ashy-headed, blue-headed, grey-headed and black-headed races in Europe and their management differs not at all. It has brilliant yellow underparts, with yellowish-green above, and a yellow eye stripe. The females of the various races are difficult to identify and the juveniles almost impossible. A very dainty bird, it measures 16.5 cm (6½ in) in length. Due to a certain amount of hybridisation which has almost certainly occurred, it can at times be hard to state categorically the race of some birds observed in the field. I once had six young birds brought to me; the sire was almost canary-like and, of the six young, some were named as one race and the rest as another; as this was by two very experienced ornithologists, this illustrates how these races can be confused.

Habitat
It is found in pasture land and open grassland, often near water but not essentially so. The nest is built in a depression in the ground where it is hidden by overhanging grass or other vegetation.

Aviary
As for the grey wagtail (page 231) but provide long grasses over half the aviary base and omit the man-made nest sites.

Food
As for the grey wagtail (page 231).

Breeding
As for the grey wagtail (page 232).

12

Woodpeckers, Wryneck, Nuthatch, Treecreeper and Wallcreeper

THESE BIRDS, ALTHOUGH GROUPED TOGETHER as climbers, vary greatly in management; as would be imagined, the needs of the diminutive treecreeper differ greatly from those of the great spotted woodpecker.

All the woodpecker family, even the close relative, the wryneck, are coarse feeders; they are long-lived when housed wisely and provided with a diet which suits their needs. The woodpeckers themselves, and here I have described only the great spotted and lesser spotted, having hand-reared only the green and black species in experimental feeding and released them when completed, can all be given the staple mixture to which one can add coarsely grated steamed ox heart and liver and grated Cheddar cheese. Give them a certain amount of soft fruit and minced vegetable matter, both root crop and greenfood. They do not seem too fond of fish in their diet, whether steamed or raw, but will take it if it is kept to the barest minimum. They enjoy hammering locusts to pieces, even the fully grown ones which, depending upon the species, may grow to 12 cm (4¾ in) long; indeed they will take everything offered in the way of livefood from the common insects of Europe.

The wryneck is better hand-reared as it will then also take everything offered. For adult specimens refer to chapter 5. When hand-rearing these birds give a reliable insectivorous mixture moistened with milk. At every other feed give livefood; if cut-up mealworms are included with the softfood, then between these meals they can have wax-moth larvae and a certain amount of wood-ant larvae but the latter must be kept to a minimum, not because of any deficiency or over-richness of the food but due to the habit of the wryneck to become too keen on it and show a tendency towards dependency upon it. Such items as earwigs and caterpillars may all be given but it is most important to include the cheese and minced steamed ox heart and liver. These birds will also take a small amount of steamed fish and roe if well stirred into the basic mixture.

Treecreepers and wallcreepers can be 'meated off' but, once again, I much prefer to obtain such birds as nestlings and hand-rear them. I would always rather bring up nestlings, when a natural rapport can be formed, than take adult birds and force them to accept me. These birds can be raised on the same diet as wrynecks, and the rule of keeping ant larvae to a minimum still applies in their case. Always resist that temptation to offer a mealworm to show how tame the birds are. Give them only chopped up in the food and be more sure of healthy stock; if cut finely they will flavour the food yet not prove addictive. Treecreepers will benefit also from a small amount of soft fruit and vegetable matter, such as brassica and dandelion, being minced and included in their diet.

The nuthatch is odd to this group in that, in the winter, as well as taking small quantities of its normal insectivorous food, it will become very dependent on its sunflower supply.

Great Spotted Woodpecker (*Dendrocopos major*)†

General Description
The variegated plumage of this bird is very beautiful and also distinctive; black is the predominant colour but the addition of white and red makes this a striking bird to view. The breast and underparts are a pale but warm mushroom shade. The throat and cheeks are white but separated beneath the cheek by the black mantle. The crown, back and wings are black and a large white patch can be clearly seen on the wings which also show a series of dots, giving the impression of wing bars. The male has a red nape and both sexes show the crimson of the under-tail coverts. In all other respects they are alike, the male being a little stouter in general appearance and the young having both crown and nape red; the central tail feathers are black and, as with the wings, the outer feathers are dotted with white. The length of this bird is 23 cm (9 in).

Habitat
Broad-leaved forests and coniferous areas will both shelter this bird and, although a special preference for coniferous trees is shown, it is still frequently heard 'drumming' in parks and orchards where well-matured trees exist. The nest is in the hole of a tree at any height between 2 m

(6½ ft) and 20 m (66 ft) from the ground; the base will be covered in wood chippings.

Aviary
In consideration of this bird's long bounding flight, it is a pity to house it in too small an enclosure. It will certainly nest in an aviary 2 m (6½ ft) high, but give a greater height where able. The length should really not be less than 6 or 8 m (19¾ or 26¼ ft) to fully appreciate their flight and provide them with adequate exercise. I would suggest, after some ghastly experiences of my own, that the enclosure be made from either railway sleepers or metal angle; if from the former, it will have only a limited life before some part will require renewal. If from metal, then one can add lengths of fallen birch or other heavier timbers for them to 'play' with, replacing them as required. These charming birds can make debris of a fine enclosure in a very short while. I would erect this bird's shelter in the actual aviary, once again suggesting such solid materials as wooden railway sleepers or telegraph poles.

Food
This is no problem: make the mixture of a crumbly consistency or a small lumpy texture, using a little honey to give the right 'feel' to it when mixing. All the additives previously mentioned can be given and the normal daily quota of mealworms chopped up and thoroughly mixed with this. Soft fruit, berries and greenfood, such as Brussels sprouts, dandelion and sowthistle heads are suitable; green peas also can be added in a minced condition and a few grated walnuts and brazils can be given when available as well as a small number of sunflower seeds daily. Some birds will ignore these and others enjoy them. This is one bird that I find does take a fair amount of peanut butter when mixed with its food, but it can be fattening and, unless the birds are housed in a large enclosure, should not be over-done.

Breeding
My birds always took great delight in taking the nest box to pieces; the bottom was first to go and there they used to roost, inside the box, their tails protruding from the hole. The only consolation was that it didn't need cleaning. These birds were very entertaining, becoming quite friendly. I learned what I could from their presence but as I could not specialise in them to the extent of providing what I conscientiously considered vital for their successful breeding, they were released. I can well imagine these birds accepting a portion of a tree bole that had previously held the nest of another of their kind.

Much patience would be needed on the part of the bird keeper who, if fortunate enough to have young result, would have some busy moments supplying the livefood diet for the nestlings. My own birds have always been hand-reared specimens, consequently learning to accept that diet which was provided, to my knowledge suitable for their needs and to them a tasty mixture readily accepted. I have little doubt that they could be reared successfully and probably hand-reared birds would be willing to give certain of the man-made mixture to their offspring.

Lesser Spotted Woodpecker (*Dendrocopos minor*)[†]

General Description
Merely the length of the hedge accentor, 14.5 cm (5¾ in), this bird may lack the red under the tail coverts but it makes up for it in its dainty appearance. The upper parts are black, heavily barred with white and the underparts are brownish-white, lightly streaked. Only the male has the red crown. The face and sides of the neck are white in both sexes and both sexes of the juvenile show the crimson cap. It is not so destructive as the great spotted woodpecker.

Habitat
A rather shy elusive little bird, it may be seen very high in the branches of conifer and broad-leaved trees, in lightly wooded areas and even orchards; the 'drumming' can be heard in spring but to a far lesser degree than that of the great spotted woodpecker.

Aviary
Although quite small, it is an extremely active bird and should therefore be given an enclosure at least 3 or 4 m^2 (9¾ or 13 ft^2), by a height of 3 m (9¾ ft). The structure, whilst not needing to be quite so fortress-like as that of the previous bird, should still be constructed of metal where possible. If using timber it is always a useful precaution to cover all wooden posts and horizontal timbers with a small-mesh wire netting. Once again, where possible, I have always had hand-reared specimens of this bird; it becomes very confiding when reared in this manner and this gives one so much greater an opportunity for study or even general pleasure in watching their behaviour. Whilst these birds are not migratory, I have always thought a shelter for their use in the colder months a useful addition to their aviary; it can be built inside the enclosure and allows them to avoid the worst weather if they so wish.

Food
Treat as for the great spotted woodpecker (page 237), but offer smaller locusts if these can be bred for their use.

Breeding
The inclusion in the aviary of suitable lengths of fallen birch will normally provide them with a nest site to their liking, but man-made boxes will be accepted; the use of the birch is purely a personal preference as I have often enjoyed watching them excavate a rotten log of birch.

In all manner of feeding the adult birds or of preparation prior to the breeding season, and of nestling feeding, treat as for the great spotted bird (page 237) but offer correspondingly smaller livefood should nestlings be hatched.

Wryneck (Jynx torquilla)[†]

General Description
Brown all over, a slim elongated bird of 16 cm (6¼ in) long, it tends to look longer due to the slender build. The plumage varies from a rich dark brown to pale greyish-brown, the background being formed of the latter with the darker shades forming the beautifully marked patterns over the whole body, wings and tail. The breast is decorated with hundreds of small crescent-like markings and the crown, mantle, back, wings and tail have their share of spotting and barring, making this one of nature's wonders of camouflage. To see this bird against the bark of a tree is a rare picture; it has a tendency to 'freeze' into immobility and to watch one if alarmed.

Habitat
Open woodland, particularly areas of broad-leaved trees, and old overgrown neglected orchard often provides just the right type of natural hole for nesting.

Aviary
To appreciate fully the beauty and grace of this species, an outdoor enclosure of at least 3 m (9¾ ft) in length by 2 m (6½ ft) both in width and height should be given to a pair. This outdoor aviary must have an adjoining draught-proof shed-type shelter attached; gentle warmth should be provided during the cold months in addition to extra light to allow feeding later in the day. Both shelter and outdoor aviary will require more vertical old tree boles, with bark intact, than horizontal

perching arrangements, although the latter should be included. I would suggest two bark-covered posts in the shelter, in opposite corners, so offering an individual roost for each bird. The outer enclosure could have a vertical bark-covered pole or tree bole in each corner and another centrally, the latter being shorter, with at least 50 cm (20 in) between the top of the post and the wire roof. This all-wire breeding area should have the four corner sections lightly roofed with rigid plastic, thus covering the corner posts or tree trunks. The most easily acquired tree boles are old rotting silver birch trees; they can easily be cut to the lengths needed when in this decayed condition. A shallow pond should be included when constructing this aviary and the base covered with a deep layer of peat or leaf mould. Little growing vegetation is needed but hops growing on one side of the wire enclosure will provide much insect life as well as forming an effective wind-break.

Food

Just as long as one gives a fairly high proportion of animal protein in the diet, the livefood can be almost disregarded for hand-reared specimens. One could not acquire an adult specimen and immediately treat it thus; certainly they can be gradually changed to accept such foods but this takes time and patience. The standard insectivorous mixture, with a minimum of fruit and vegetable matter, generous servings of grated cheese, steamed and minced ox heart and liver, and small quantities of fish and fish roe treated in a like manner will be suitable. Any livefood should be cut very small and mixed well into the food. When wood-ant larvae are obtainable take as many as can be gathered and freeze them for all-year-round feeding; not much at a time will be required – a pinch per bird mixed into the food will suffice. Greenfly can be shaken over their food dish and will be readily taken.

Breeding

I cannot resist allowing a personal preference for this bird to show itself here. It is a bird which has always fascinated me since the time as a child when my father pointed out a nest in a nearby orchard, situated in the hole of an old apple tree; it contained eggs at the time. I was lifted shoulder-high to have a quick peep and nearly fell backwards when the brooding hen hissed snake-like at me and darted her head forward fearlessly. Since those days I have hand-reared a number of nestlings and 'meated off' many adult specimens. Where a pair have not nested within a certain time, or when a single bird was being held, I have liberated them, readily confessing a sentimental fondness for this bird.

One of my biggest thrills in bird keeping was to watch a pair of these birds go through their mating display. Sitting on a perch facing each other, they raised their heads high, crests raised and bills open; they shook their heads in nodding fashion up and down for minutes on end, only stopping when they realised that I was watching them. That year I had great hopes of breeding them but an attack of botulism that affected many birds put paid to this. I have experimented a great deal over the years with various diets for this bird and have a greater confidence now than ever before; perhaps one year I may be fortunate again. One problem is to identify a true pair. In the many that I have studied, each loud-calling specimen was the male, and each of these was just a little more heavily built. Were these birds given a section of tree which held a natural site, they might well, as mine did, enter and accept this; I had placed decayed wood chippings in the hole in readiness for their inspection.

The rearing of nestlings, if one was ever fortunate enough to have some hatch, would be a challenge, but I see no reason why success should not result. Our failing in general, as bird keepers and would-be breeders, is to keep too many species; the breeding time stretches to the limit our insect production. Specialisation in a limited number of species is a more likely recipe for success.

Nuthatch (*Sitta europaea*)[†****]

General Description
It is bluish-grey above, on the crown, mantle, back and tail, and the chest should be a rich warm timber-brown. The flanks of the male are a bright reddish-chestnut hue and the ventral area is heavily spotted with the same colour. The female has far less of the chestnut on the flanks and not so deep chestnut flecks over the vent. The black stripe through the eye and the whitish throat, indeed the general colouring, make them an attractive subject for a garden aviary. The length is 12 cm (4¾ in).

Habitat
High in the branches of trees, broad-leaved being the favourite haunts, it can be seen anywhere. It feeds lower down in the winter and will then often visit gardens and take the foods hung up for the tit family. It nests in a natural hole in a tree and will normally mud-up the entrance to allow just sufficient space for its own entry.

Aviary

Allow them an enclosure at least 3 m (9¾ ft) long by 1.5 m (4⅞ ft) wide by 2 m (6½ ft) in height. Secure two or three log-type nest boxes around the aviary. A shallow pond is always welcome since they enjoy bathing as much as any bird. The remainder of the aviary floor space can be evergreen shrubs and flowering decorative bushes; a few ferns and, for attracting insect life, the old favourite climber, hops, may be planted at the rear. These birds will winter outdoors quite well but should have a small roofed section at each end of their enclosure to roost beneath and keep free of severe frosts.

Food

One of the simplest of insectivorous birds to cater for, provide it with the usual mixture, with all the normal recommended additives. During the winter months this can all be supplied in very small quantities as the nuthatch becomes almost a seedeater at this time; sunflower seed will then form its main diet, together with the cut-up mealworms and adhering particles of softfood, cheese etc., of which it is extremely fond.

Breeding

This was one of my first-ever insectivorous species as a child and they still give me as much pleasure to watch today.

A most entertaining bird during the breeding season, the male becomes quite an ardent wooer, calling and fetching livefood for his mate from dawn to dusk. As the female becomes aware of what is expected of her, she will increase her visits to one of the sites, often having to follow her mate inside to obtain the proffered insect. The nest, if one can call it such, consists of dead leaves, small thin particles of bark and rotting wood chippings, all forming a shallow bowl in which she will lay her eggs.

Here again we have a species which, if hand-reared, will pass on to nestlings a certain amount of softfood; the provision of extra mealworms, cut into small pieces and well mixed, will add their flavouring to the food and encourage its consumption. Much livefood, in the form of spiders, earwigs and aphids, will be collected from a climber such as the hop if its growth is encouraged. Other insects can include maggots, once they have been cleaned properly and treated with the additives. Offer almost anything at this time: it seems that very little is ever refused. Give small locusts at first but, if the size is gradually increased, fully grown specimens can eventually be offered and they will hammer them to pieces for their constantly hungry and calling young. The wax-moth larvae and other smaller, softer items are ideal for the bare nestlings just hatched.

Treecreeper (*Certhia familiaris*)[†****]

General Description
A brown bird of 12.5 cm (5 in) length, attractively marked, rather like a wryneck in pattern, the plumage is palish-brown on the upper parts and darkly dotted and streaked in very evenly marked patterns. The throat, breast and belly are white and the bill is slender and slightly curved. The tail is brown.

Habitat
Of woodland areas, frequenting both broad-leaved and coniferous trees, it often roosts in bark cavities of red cedar where this exists. They nest in a cavity behind raised bark, a natural tree crevice or even behind bark and ivy, wherever there is space to hold their nest of grass, rootlets, moss, wool and feathers.

Aviary
The standard minimum recommended size of enclosure will be adequate to house a pair of these birds. It is advisable to add to this a frost-proof shelter, which, although it may not be heated, can provide extra feeding time if electricity is laid on for lighting. There is widespread destruction of elm trees due to the Dutch elm disease and this makes the collection of suitable large 'sheets' of bark a simple matter; these can be firmly attached to a dry rear wall of the aviary to form one or more natural nesting sites. The inclusion in the aviary of a few old rotting birch-tree boles will give these birds a near-wild setting and provide quite a lot of insect life. The growth of a climbing plant, such as the hop, will allow them to bathe in the foliage in their natural manner, although these birds will also use a shallow bath.

Food
A very fine mixture is to be supplied with the cheese grated small, the steamed ox heart and similar all minced very fine and four or five mealworms per bird daily, chopped finely and well mixed in. Moth larvae can be given alive but also stirred into the food. If obtainable give a pinch of defrosted ants' 'eggs' from a frozen source; the dried variety offer little more than roughage, and even when soaked first in milk containing Abidec (Parke, Davis), they introduce only what one can quite easily get

the bird to take in other simpler ways. Fruit can be kept to a minimum but greenfood should be minced finely and added daily; try to offer this in variety, not always giving the same vegetable matter.

Breeding
Breeding these birds for the first time ever gave me great satisfaction. I had experimented with various foods over a number of years, having been granted licences for research work and I had on occasions reared from the day after hatching. It is a time-consuming task and one must ensure warmth at night, this being the vital ingredient to success. To rear birds on a food mixture, and to keep them relying upon it before attempting to breed from them, does help to a certain extent because they are already accepting it as a basic food. I am positive that my pair used some of this mixture daily when rearing their own nestlings. We did, of course, supply quite large amounts of moth larvae in particular and the mealworm content of the softfood was greatly increased too.

Wallcreeper (*Tichodroma murina*)[†]

General Description
An exceedingly attractive specimen, the crown and mantle are bluish-grey and the wings are dark brownish-grey but with very large crimson red patches extending from wing butt almost to the tip, the edges being dark and white-spotted. The tail too is very dark with just the outer feathers tipped white. The throat during the breeding season is black like the breast and belly but, during the winter months, it is a very pale greyish-white. The sexes are very alike but on close inspection one finds the female to be just a little less bright in the crimson and the beak of the male to be a fraction longer. With such slight sex differences one must have two or three birds for comparison. The flight is rather butterfly-like, resembling that of the hoopoe. The bill of the nestling is much shorter, not growing to its full length until it reaches independence. The over-all length of adult birds is 16 cm (6¼ in).

Habitat
A species of mountainous southern Europe, it rarely perches on a tree but uses steep rock faces, and gorges where crevices abound, replacing the rock nuthatch at a certain altitude. It nests in rock holes and cavities.

Aviary
Because of its manner of flight, and in order to appreciate the beauty of the bird climbing, I would suggest an enclosure at least 4 m^2 (13 ft^2), with a height of 2.5 m (8⅛ ft) being the minimum. If the aviary walls are constructed to these suggested dimensions, giving an area of 10 m^2 (33 ft^2) per wall, a natural ornamental rock-faced wall could be built on one, two or three sides; alternatively, using the usual wire mesh, a central large rock pillar-type block could be built in the middle of the aviary, allowing all-round flight by the birds. Whichever type of aviary is constructed it must be for the general well-being of the inmates and it should always be remembered that this bird rarely uses a normal perch, preferring, as does the nuthatch, a rough surface to climb. When building a rock face in this manner, it will be found far simpler to first erect an ordinary brick wall into which the ornamental rock can be built when it is fully dry; cavities for roosting and nesting can then easily be incorporated in such a structure. These birds do appreciate bathing facilities, and I mention this because, with a little planning, a very attractive setting could be provided for the inmates, with shallow pools at different heights and pumped water cascading from one to the other; advance planning of this nature can make a tremendous difference in providing a near-natural habitat and appreciably raises the standard of aviculture. Extra hours of light, for additional feeding time, in a frost-free shelter, should not be denied this bird, although artificial heat is not vital.

Food
As it ranges almost to the snow-line high in the mountains, the food taken by this species in the wild might be imagined as rather on the coarse side. However, at that height there are many types of insects, even butterflies, and the birds spend most of their time darting, with nuthatch-like movements, to crevice after crevice in search of some hidden insect in between the rocks. On the standard insectivorous mixture as previously advised for the treecreeper (page 243), they will live for a number of years. If an ivy can be grown in their aviary and trained up a rock face it will encourage many insects; even so, one must tempt these birds to visit the food dish frequently, otherwise they tend to spend all their time searching. The addition of mealworms, moth larvae and small locusts, all cut into small pieces and mixed into the food, will normally ensure a healthy daily food intake.

Breeding

The nests that I have seen in the wild were all in rock holes and crevices and constructed of grasses, moss, rootlets, some animal wool and hair and then lined with feathers. I have never had such a bird nest in my possession, the majority of the birds that I have handled being purely for a study of dietetics. The near relative, the treecreeper, nested when I was carrying out feeding experiments and I confess that I really had to do very little to assist them.

13

Starlings, Waxwing, Bee-eater, Golden Oriole, Hoopoe, Roller, Kingfisher and Shrikes

THOUGH GROUPING THESE BIRDS by size, some require specialised treatment whilst others can be very coarse feeders.

The starlings will certainly exist on an austere diet but, being a firm believer that 'we are what we eat', and having seen the excellent colours which result, I could never condemn this bird to anything less than good feeding.

Waxwings again have a very hard time in the wild during some winters, not just in the British Isles but in many countries when a natural harvest of berries fails. These birds will keep in wonderful condition if given, as well as the basic insectivorous mixture, soaked household currants and a piece of eating apple. In the basic mixture should be grated plenty of vegetable matter, brassica and dandelion.

Bee-eaters and hoopoes need plenty of animal protein. Build up their basic softfood with extras such as steamed and minced ox heart and liver, and fish or fish roe, treated in a like manner. They will also benefit from eating a small amount of vegetable matter daily, plus the usual locusts, mealworms and moth larvae, which should be cut and mixed well into the food.

The oriole and roller, whilst sharing this preference for live insects, should also have a few soaked currants and grated fruit, with the normal daily additions of grated cheese, ox heart and other protein food. These larger birds will also take live locusts but, although it is pleasant to observe birds feeding naturally, it is far better for the bird in the long run, for longevity and general fitness, to have all such livefood cut up and mixed with their food.

Starling (*Sturnus vulgaris*)[1*****‡]

General Description
In case any whetting of the interest is needed for this bird, let me state immediately that the tame starling is very different from the wild bird. Especially if hand-reared, the plumage itself is a mass of colours, high-lighted when hit by the sun, and the daily bathing, better food and clean roosting site all add up to making this a more colourful specimen than its wild kin. At 21 cm (8¼ in), this bird is largish and therefore shows a greater expanse of colour to advantage; it is blackish at first glance maybe, but the iridescent hues of green and purple with the spangled spots make this bird quite handsome. If hand-reared and housed near a good songbird when they become self-supporting, they will learn to mimic extremely well; we had one male that imitated a goat so well that people used to ask who had a goat nearby. It makes a good parent and I have used one as a foster parent for other birds, one being the rose-coloured starling (*Sturnus roseus*), which a friend brought to me as eggs to place under these birds. It is also a fascinating bird to study.

Habitat
It frequents gardens, woods, cultivated areas, villages and towns throughout Europe.

Aviary
The minimum-sized enclosure is suitable for a pair but these birds have a habit of breeding, so unless wishing to change later, it is advisable to start off with as large an enclosure as one can manage. Bathing facilities are essential and the amusing bathing habits can honestly provide one of the best reasons for keeping these birds. Make sure the aviary floor is a well-drained one. These birds will turn over the soil better than any cultivator. Give a few large nesting boxes; some individuals will roost in them, others will choose the open. Roof over a small portion of the enclosure at each end for comfort.

Food
I am well aware that one can keep these birds on scraps, but when providing the usual diet of insectivorous mixture, plus all the usual additives recommended, the gloss on the plumage and hues which reflect the light are all improved. Quite apart from this, if hand-reared, they

remain remarkably tame and confiding, very mercenary really, and will go to most any lengths to obtain mealworms from their keeper.

Breeding
With the provision of nest boxes, and a bundle of materials hanging up so that one can watch their antics, they will soon go to nest. The sexes look very alike at first glance but the male is the brighter of the two. Also, when they are in breeding condition, study the bases of their bills: the point where the bill joins the head is blue in the male and pink in the female.

When rearing is in progress, they will need the usual multitude of insects. Locusts of all sizes are wonderful for starlings; also provide cut-up mealworms in a fair number in their food. Wax-moth larvae just seem to disappear as soon as they are seen. They will rear on worms and I have seen many acquire various complaints from a diet of these so they should be watched carefully and made to revert to softfood if trouble occurs. This is one bird which will give its young vast amounts of softfood and rear them successfully on it.

Rose-coloured Starling (*Sturnus roseus*) ****

General Description
This is of the same size as the common starling, 21.5 cm (8¼ in) in length. Although the head and throat, wings, rump and tail are all black, the mantle, chest and belly are pink. The sexes are similar but the female is the paler. Juveniles are very much like the young of the common starling. These birds are crested, but this is not always noticeable as the crest is not held in the raised position at all times. The female has a slightly shorter crest than her mate.

Habitat
It belongs to the former USSR and eastern Europe, dwelling on the grassy steppes and low surrounding hills where it breeds in colonies. It nests in rock crevices.

Aviary
As for the common starling (page 248). Our birds were starling-reared.

Food
As for the common starling (page 248).

Breeding
As for the common starling (page 249).

Waxwing (*Bombycilla garrulus*) ****

General Description
The smooth greyish-brown plumage of this bird must be touched to appreciate fully the softness which exists in a number of the northern varieties; it is almost comparable to the feel of goosedown. It has a rich chestnut head crest, pale underparts of grey, slightly tinged with pink and a black throat. The lower back and rump are grey and red, yellow and white markings on the wing show up to perfection when close by. The red is of a coral hue and looks for all the world as if it is composed of wax. The tail is tipped with yellow, the legs and bill are black and there is an eye stripe of the same colour. The length is 23 cm (9 in) and, seen as a silhouette against the sky when perching, it can resemble a starling until one sees the crest. The sexes are very alike, but upon close observation one will see a longer and wider crest in the male, and this may be perhaps a shade darker. The black throat bib in the male is wide and deep, ending very abruptly, whereas that of the female is narrower and tends to fade out on the upper breast. The yellow 'tick' marks in the wing also differ: both the downward stroke of the tick and the upward 'cast-off' are far thicker in the male than the female.

Habitat
In the birch forests and tundra areas of Scandinavia and the former USSR, it may nest colony-style in one place one year and a hundred kilometres away the next.

Aviary
Give a pair an enclosure measuring 4 m (13 ft) long by 2 m (6½ ft) wide and high. Place perches at each end and put their food in the centre but shielded from strong sun or rain. Each end of the aviary should be roofed for rather more than 50 cm (20 in) and have a small amount of side shelter to prevent driving rain reaching a roosting bird. No other winter quarters are necessary and, provided that they are fed correctly, they should live for many years.

I once had a specimen that was about 15 years old and still as fit as ever. Site the aviary in a cool spot; it should not be too sunny.

Food

Give daily a handful of soaked household currants, previously steeped in water for 24 hours; during summer keep these in a refrigerator to prevent any fermentation due to heat. Sweet apple should always be available, with a small quantity of insectivorous mixture into which has been mixed a little grated carrot, cheese, brassica and about two cut-up mealworms per bird. Never at any time feed worms or live mealworms to this bird. Maggots will often pass straight through this species without being digested. It raises its young in the wild on mosquitoes and cranefly. Under controlled conditions, the first time they were ever bred we managed to obtain large amounts of wood-ants' larvae; one hen in particular was so tame that she took these from a moistened finger, raised herself and regurgitated them to the newly hatched young. From this point we had little difficulty; within an hour or so they were using wax-moth larvae and ant larvae and continued with these foods, along with softfood, until the young were about 7 or 8 days old.

Breeding

Since they nest on the colony system in the wild, there is nothing to prevent one attempting this method of breeding them under controlled conditions. In fact, at one stage, I had seven hens all with nests at the one time and there were about five males in the enclosure, but this was a well-proportioned aviary.

The nesting sites may range anywhere between 1 m (3¼ ft) and 2 m (6½ ft) from the ground. My birds have nested at all heights within that range and wild specimens frequently have to nest this low because of the stunted trees in the tundra areas. Shelves, clusters of branches and shallow boxes will be used when a female is in breeding condition. At this time the problem is to make them choose a safe site, rather than one in the open or in full view of owls or cats; they have little fear at any time and, with breeding condition, they assume the tameness expected of northern species. Provide grasses and moss – mine often used dry chickweed and other plant stems, animal hair and vegetable down – that from sowthistle is frequently used; a few feathers and wool will also be used for the lining.

These birds are late fliers and at dusk they will often hunt moths or small insects and probably mosquitoes – the latter especially as we have always been in the habit of having a pond in their enclosure. They will at this time fly from one end to the other continuously calling their low 'si – i – i – t – si – i – i – t', which is a low trilling sound.

Their mating behaviour is most interesting and their gift-offering procedure can be fascinating to observe. They will select an item, anything from food to a small piece of twig or leaf, then pass this back and forth between themselves; this can go on for many minutes at a time and all the while they hold a grotesque shape, puffing their feathers from the body like a large ball, raising the head and pointing the tail vertically downwards giving themselves a very exaggerated size and shape. I would say without hesitation that, up to the present time, I have not found better rearing items than wood-ants' or wax-moth larvae for these birds; day-old locusts will be taken but they disappear very quickly in a large aviary, especially if one has a colony of these birds on the look-out for insects. The larvae recommended can usually be obtained or cultured in such vast numbers that one can freely offer them at frequent intervals.

Bee-eater (*Merops apiaster*)[†]

General Description

Once this bird has been hand-reared one does not forget it easily, not merely because of the vivid hues of its plumage but also because of the mannerisms peculiar to this bird: the enormous gape; the habit of almost swallowing one's fingers if left in the close proximity of the bill; the greed with which they bolt their food at the nestling stage. Despite these failings in etiquette, it grows into an attractive bird of harlequin plumage: the greater part is blue-green whilst the head and mantle are chestnut, the latter fading on the back to the yellow rump. The throat is yellow and the forehead yellowish-green; there is a dark, almost black, eye stripe and, below the yellow throat, a black breast band which separates the throat coloration from the blue of the breast. The shoulder is dark greenish, there is a chestnut patch on the wing and the flights are blue with a black trailing edge. The tail is greenish-blue with the central feathers projecting to two separate 'swallow-tail' points showing clearly against the shorter remaining tail feathers, which spread on each side. The length is 28 cm (11 in) and the sexes are very similar. I feel that the hen is a shade duller of plumage and perhaps a fraction shorter in the bill; she is certainly the quieter, or rather, the less noisy, when being hand-reared.

Habitat
A bird of southern Europe, mainly Italy and Spain, this bird colony breeds using holes in the banks of streams, sand pits or similar straight-faced, sandy-soiled terrain.

Aviary
Having considered its habitat, one should give much thought to the shelter to be used during the winter months. This should adjoin the outdoor breeding aviary and will need to be well glazed, preferably from above to allow maximum light to enter; it should have both heating and lighting facilities. Although great warmth is not necessary, a gentle warming of the air is a good thing; artificial lighting will have to provide extra feeding time for these colder months. I would suggest that these birds be allowed the use of the outdoor aviary all the year round, but be able to return to the warmth of the shelter when wishing to do so; a well-illuminated shed will draw them inside, even if only to eat from their food dish and this should always be sited indoors. The outdoor aviary should be 3 or 4 m (9¼ or 13 ft) long, by at least 2 m (6½ ft) wide and high. Where the aviary joins the shelter, the roof should be covered for the space of 1 m (3¼ ft) with a clear plastic building material; beneath, in the dry, site a large bale of peat which should have a few holes made in it at the sides near the top. If it can be managed, I would suggest this bale be stood on a wooden platform in an effort to keep it vermin-free. The holes should be the diameter of a fist. If these are not to the birds' liking they will amend them to suit their needs.

Food
A friendly apiarist will now be needed: ask for any wasted brood comb, or try to obtain a few wasp nests (see page 32). Some wasp nests have been treated with chemicals and so could be useless but others may in any case be safe if all the grubs, which we now wish to obtain, are sealed in their own little chamber. With a pair of tweezers one can soon draw out these grubs for immediate use or for freezing for later use. These items are extra to our basic softbill food which should receive liberal quantities of the many nutritious additives, plus a little more honey. Mealworms can also be given, a dozen a day per bird, cut into small pieces and mixed well with the other foods. The use of locusts and wax-moths and their larvae should be watched carefully; ensure that the birds are taking softfood and, if they seem to drift from it, cut down on the live insects and increase those cut up and mixed with the food. One cannot of course use such persuasive methods when

the birds have young; then they must have a maximum supply of all possible.

Breeding
The beginnings of the tunnels in the bale of peat would be sufficient for nest sites if a pair of these birds reach breeding condition: they will then excavate this further to the shape and angle they require. No nesting material is needed. The entrances should be started in the end of the bale so that the nest chamber can be situated at a maximum distance from the entrance itself.

Hand-reared specimens are more likely to rear than wild adults taken from liberty; they should also keep to their rearing food to a certain extent and subsequently be less reluctant to offer it at an early stage to any nestlings one is fortunate enough to breed. So many foods can now be stored by freezing that one can build up supplies to quite large stocks in preparation for such events as breeding. It is then a simple matter to de-frost them as required, add to the softfood and offer them, together with any live insect specimens.

Golden Oriole (*Oriolus oriolus*)[†]

General Description
A bird of 24 cm (9½ in) long, the male is a bright yellow with wings and tail of black but showing yellow on the tail's outer feathers; from the bill to the eye it has a small black eye-patch. The female is considerably duller in appearance, being of a greener hue and only yellowish on the belly. The juveniles are darker still and more dull of plumage. In captivity, without colour feeding, it tends to lose some of the sheen despite every effort to maintain the bloom on the plumage.

Habitat
Watching this migrant in its native haunts, southern Europe's broad-leaved woodlands, forests and parks, where there are scattered trees, is a thrilling experience. The nest in the wild is a work of art, suspended between the branches of a tree.

Aviary
I have had only one pair of these birds; these were obtained from abroad and were being kept until their owner could collect them. I would suggest that a 5 m (16½ ft) long enclosure would not be too generous; the height should be above average, perhaps 3 m (9¾ ft) and the width the same,

but one must realise that this bird, if ever obtained, is something very special in aviculture. A real effort would have to be made to breed from them even to justify having them in the first place.

Food
In addition to the basic food supply of home-made softfood, one should include locusts in this bird's diet. It may well be that fresh vegetable matter assists the coloration and we should certainly increase the amount of greenfood and give plenty of soft fruit. The usual mealworms should be added daily and, despite this bird's size, such items as wax-moths and even smaller insects, are readily taken.

Breeding
I would think that the inclusion of two small apple trees in the aviary would assist one to specialise in the study of this bird and do all possible in persuading it to breed. The nest is slung between two branches; the rim is of woven twigs and vegetable stalks whilst the nest itself is of grass, plant stems and pithy bark, all interwoven, binding the nest to the rim. The lining of the one I investigated was of fine grass, vegetable down and seeding grasses; quite possibly the materials could vary with the locality.

Any young of this species could be given fairly large insects from hatchings, or newly hatched locusts, cut-up mealworms and wax-moth larvae; even small fish could be tried. I have often been surprised to see various, even far smaller, birds, take minnow fry from an aviary stream and have seen some young birds reared on them almost entirely. I have also seen many take tadpoles and small frogs.

Hoopoe (*Upupa epops*)[†***]

General Description
Unmistakable for any other bird, the hoopoe has, with adulthood, a long curved bill and a crest which, although often relaxed in the flattened position, can be raised in moments when curious, alarmed or displaying. This, when erected as a fan, shows black tips and a white bar between these and the general pinkish-beige colour of the lower crest. The head, neck and body, wings and tail are barred black and white. In flight this bird really does look like a very large fluttering butterfly. The female is very like the male but slightly smaller in stature; the bill, too, is just a shade shorter and the crest not quite so prominent. These identifying features must be seen when several birds can be compared; the sexing

of a single bird can be difficult. The main call, in other than the breeding season, is 'hoop – hoop' with an extending of the throat; with the completion of the call, the air seems to be expelled with a sound of 'kh-u-u-u-r'. When the male reaches breeding condition he seems to add an extra 'hoop', making a series of three in his call, and the exhaling sound also becomes louder. Two of our hens laid at opposite ends of a long horizontal nest box and the one male present could be heard half a kilometre away so loud was his call at that time.

Habitat
Orchards, vineyards and wooded areas are its haunts. It is a hole-nester, generally using a tree, or hole in a building wall.

Aviary
The enclosure we gave our birds, which were all hand-reared specimens, was about 8 m (26¼ ft) long. Prior to this we had, at one time, eight young which we allowed out of their hand-rearing box to fly around indoors. They are a fascinating bird whilst young; their bills are far shorter but these gradually grow so that, when ready for liberation into an outdoor enclosure, they are almost full length. Bearing in mind their manner of flight, one should endeavour to give them as large an aviary as can be managed. Some frost-free winter quarters are advisable but we allowed our birds the use of outdoor flights all the year round. A little warmth may be necessary but extra feeding time is certainly necessary for complete fitness. The beauty of their flight often drew me to their enclosure after dusk to see them.

Food
Although many aviculturists say that these birds do not drink, and, having never seen them drink, I believe they may get all their required moisture from their food supply, I did at all times see that they had bathing facilities in their aviary; despite this they continued to dust-bathe as always.

The hoopoe throws back its food in a manner similar to a toucan or hornbill; holding the required item between the tips of the bill, it will toss its head back, at the same time releasing the morsel of food so that it arrives at the back of the throat as the head reaches its furthest point back. Remembering this, it will be seen that any softfood taken must be anything but fine; it should be rather lumpy. We gave the usual softfood with all additives such as ox heart, liver and fish with very little fruit, but kept to the usual quantity of vegetable matter. We also gave a very little finely minced beef and fish, both in the raw state; a little honey was also added and the whole, together with about six cut-up mealworms per bird, became a rather lumpy mixture. Pieces the size of a pea were very easily thrown back by the young birds and this carried on throughout their adult lives.

Breeding
Our birds did not use any nesting materials on any occasion. The nest boxes all contained a few centimetres of peat, slightly moistened beforehand and the birds merely formed their own depression and laid in this.

For rearing purposes, offer locusts of all sizes; whilst nestlings are very small give newly hatched ones, together with any other small insects available. With the pond in these birds' aviary they had a large number of tadpoles and small frogs; we saw them take these on many occasions, much to my wife's dismay, but apparently to the delight of the hoopoes.

Roller (*Coracias garrulus*)[†]

General Description
The roller is very conspicuous in flight or perching, the brilliant plumage immediately identifying it from all others. The head, nape, breast and underparts are blue and the tail is blue with black tips; the central feathers of the tail are a much darker blue. The wings are also dark blue, tipped and edged with black, and a large bright but paler blue wing patch is very prominent; the back is chestnut. The overall length is 31 cm (12¼ in). The sexes are very alike but, when compared side by side, the female is slightly less bright of plumage and also slimmer. The young are more dull and remind one more of the Indian roller which is far browner in appearance.

Habitat
This migrant likes open country with scattered trees, and usually nests in a tree hole or in a building wall, although it is found also in rock crevices. It breeds in eastern and southern Europe.

Aviary
The aerial acrobatics of this bird in the wild cannot be performed fully in the normal aviary. Due to its size alone, however, it should be given as large an enclosure as possible; I would suggest that a breeding pair would need an aviary at least 8 m (26¼ ft) long by 4 m (13 ft) both high and wide. These birds will winter outdoors but a draught-proof

shelter should be given them, where they can avoid the worst of the frosts and feed in comfort.

Food
All of the birds I have kept have been hand-reared. We used locusts, mealworms and our own insectivorous mixture moistened with milk; steamed fish was mixed into the food at first, and later added raw but minced. Later, when adult, their food varied little but was given drier, of a more crumbly nature, and with added maggots.

Breeding
Although never having attempted to breed from this species, I have seen a number of nest sites; most had a certain amount of old plant stalk, feathers and the usual debris one finds in the case of such hole-nesters. There is no true nest. A large box with a suitably sized entrance hole, would, I imagine, be acceptable to these birds; the floor of this should be about 8 cm (3⅓ in) deep with a mixture of peat and dampened sawdust or wood shavings.

For rearing one could offer all the usual rearing foods and, due to the appetite of larger types of birds, I would suggest maggots be made use of after having been treated with a few drops of cod-liver oil, and various powders made up from calcium lactate, Vionate (Squibbs), Casilan and Complan (Glaxo).

Kingfisher (*Alcedo atthis*)[†]

General Description
Apart from the bee-eater, this is the most colourful British bird; in the sun the upper parts appear a brilliant mass of cobalt-blue and green metallic hues, beneath which is rich chestnut. The throat is white and a patch of white is very conspicuous each side of the neck. A chestnut stripe runs through the eye and, just in front of the eye is a small white splash which, even though quite minute, is easily seen when near the bird. The length is 16 cm (6¼ in).

Habitat
It may be found along clean freshwater streams and lakes where fish abound.

Aviary

I have kept only one family of young kingfishers: these I hand-reared while still at school and I vowed then never to do so again. A suitable enclosure for a pair would have to be 6 or 7 m (19¾ or 23 ft) long and at least 2 m (6½ ft) in height and width. We can consider keeping these birds on a natural or on a home-prepared diet. If one were able to collect discarded fry from a trout-breeding establishment or obtain small fish in any other way, in large and regular quantities, the former would be well worth considering; it is useless and cruel, however, to embark on any exercise like this if there is any likelihood of supplies failing half-way through a programme. I mention this here because, were one keeping these birds under very natural conditions, they should have a pond in the aviary, shallow but long, covering about half the area. If keeping them on a basic home-made preparation, then bathing facilities only are needed. An open-fronted shelter will be necessary if they are being fed on a softfood with a variety of additives but a closed-shed shelter if they are being fed indoors during the winter on small fish. This shelter would again have to have some form of shallow indoor pool and very roomy quarters.

Food

When hand-rearing a bird it is very easy to give food which would otherwise rarely if ever be eaten; it is a simple matter to persuade a young bird to eat a man-made diet, to grow up on this and so accept it in later life as a natural course.

The five kingfishers which I had were dug from a bank during the excavation and erection of a riverside building not so very far from where the Magna Carta was signed. I took them home and we were fortunate that, at that time, some damming of a minor stream was going on, with the diversion of the water into another. This meant that the lower reaches were mere mud and shallow pools in which we could gather hundreds of small fish; the pools were so thick with them that a dip into the moving mass with a net meant the capture, literally, of hundreds and buckets were soon filled in this manner.

The first time I gave a feed of freshly killed fish, one bird passed its excreta, expelled as was normal, but at that time totally unexpected by me, for a distance of about 50 cm (20 in); I queried it with my father as I thought the bird was ill or that something was wrong with my feeding method. The smell was not too pleasant and my mother's carpet not exactly the way she most appreciated it. I suggested returning them to the site but my father replied with, 'You took them, now you look after them.' We found that, by rolling up a sheet of paper and attaching this to the side of their now three-sided box, they turned each time with their backs to this make-do tunnel and all was well if one held one's breath and concentrated deeply upon the end being fed; my mother's trance-like expression persisted for many days but fortunately my father's sense of humour and appreciation of birds won the day. The birds thrived despite a young lad pushing mealworms, fish and soft pellets of a food mixture down their throats. Every member of my family had a rather pained look about them when I brought out the birds for a feed; most would suddenly remember an urgent job elsewhere, or think it time to take a walk in the garden among the lavender and roses.

With the increase in size came appetites to match. I was young but can well recall the food mixture prepared by my father for all our softbills in those days; these birds ate the very same but had more mealworms, grasshoppers, minnows, sticklebacks and minced larger fish than any others. The pellets were being thrown up as fast as I could manage to keep the place clean. I do not know how many fresh, new, clean boxes those birds had. The disapproving looks of all pursued me everywhere; the 'go-and-wash-yourself' instructions came far more often. I cared little at the time because these birds, despite their more dull juvenile colours, were in wonderful condition; I recall well that when my father gave me a 'well done' they must have really warranted such a congratulation. Their bills grew fast and until we liberated them were still completely black without the orange on the base of the lower mandible.

Eventually, despite their quite high food intake of the basic softbill mixture, we began to give them live fish in an old hip-bath. They would all perch around the rim, being housed now outdoors in a small 2 m (6½ ft) long aviary, and the heads would each point at a fish; suddenly, as though it had been fishing all its life, one would make a dive and hammer the fish on the metal bath rim. The next bird, in such circumstances, would first beg, then snatch. When all were feeding well they were liberated, but up to that time they had still been eating large amounts of softfood containing minced fish and cut-up mealworms.

Breeding
Although never having attempted to breed from this species, because of being unable to provide suitable accommodation (the ideal would include natural running water), I imagine that with hand-reared specimens, it need not be impossible if one perseveres and specialises in this effort. Few if any species are impossible to breed if first studied thoroughly then persevered with – this has always been my contention and why I have achieved so many first breedings over the years. Where even

pumped water forms a stream and sufficient fish fry can be obtained – perhaps the discarded fry from a commercial trout farm – the 'fabricated bank' of one or more bales of peat, well holed, might well provide a suitable nesting site. My own specimens, whilst I kept them, took softfood containing extra fish meal, fish roe, minced fish, cut-up mealworms and a number of other insect items as acquired; I see no reason why, with a regular and plentiful supply of minnows etc. in the stream, breeding should not be achieved. However, it is certainly not an ambition to be taken lightly.

Great Grey Shrike (*Lanius excubitor*)[†]

General Description
The largest of the shrikes, at 24 cm (9½ in) long, the forehead, crown and mantle are a dark pearl-grey, the throat is white and the chest is pale grey, fading to white at the vent. A black eye stripe runs from the lores to the ear coverts and this is very prominent, as are the black wings and tail; the wings have a white stripe. The female is duller and the young even more so; both have slightly barred breasts, though that of the young is more heavily marked.

Habitat
It is found in wooded country, orchards and open forest, normally nesting in thorny bushes or in trees.

Aviary
As it has a strong long flight, rarely settling on the ground, the aviary should be 6 or 7 m (19¾ or 23 ft) long and 2 m (6½ ft) in height and width. At the rear an open-fronted shelter will provide all the necessary seclusion and winter warmth. If one can have a dense hawthorn hedge immediately outside this shelter, but to one side so that it does not limit their flight, it will form a more or less natural nesting site; a little clear plastic roofing over this could prove an advantage. The ground cover should be short and fairly bare; provide a small pond for bathing. The only specimens of this species which I have kept have been hand-reared birds and so were tame. Having handled wild ones abroad, I somehow cannot see a wild bird settling sufficiently to breed in an aviary.

Food
Due to my birds having been reared by hand, I have always been able to feed them a basic softbill mixture, with a maximum of animal protein

added; in addition we have given full-sized locusts and small mice. They are simple to cater for.

Breeding
Although I have kept about a dozen specimens over the years, I have never attempted to breed from this bird. Certainly much livefood would be required but most of this could be in the form of young mice especially reared for the purpose. The birds always emptied their softfood dish and they readily accept this food if they have been reared upon it.

Provide dry grasses, twigs, rootlets, plant stalks, animal hair, tufts of wool, teased out finely to avoid accidents, and finally feathers. I lost a hen once, through her becoming entangled with sheep's wool, and apart from the regret at a loss through carelessness, it is extremely frustrating to raise a bird to reach this peak condition, only to see it die.

Red-backed Shrike (*Lanius collurio*)[*]

General Description
A handsome bird, the male has a blue-grey head and rump, a rich chestnut back and a long graduated tail of black with white feathers at the side. A wide black stripe runs through the eye from lores to ear coverts and the throat is white, becoming a pinkish-cream on the breast and underparts. The female lacks the facial stripe and is mainly brown. Juveniles and the female share small crescent markings on the underparts. The adult length is 17 cm (6¾ in).

Habitat
This migrant frequents scrub and heathland with scattered trees, frequently nesting in hawthorn or shrub.

Aviary
In a small aviary one would lose much of this bird's natural beauty, particularly of the flight and the hovering when spotting a morsel of food on the ground. It rarely stays on the ground for long. Provide an enclosure similar to that for the great grey shrike (page 261), but remember this bird should either have a frost-free shelter or be moved into an indoor flight during the winter when extra feeding time will be greatly appreciated.

Food
As for the great grey shrike (page 261).

Breeding
As for the great grey shrike (page 262).

Woodchat Shrike (*Lanius senator*)†***

General Description
The male is a most striking specimen when in full breeding plumage, with a rich chestnut crown and nape and a black forehead and cheeks. The underparts, rump and wing bars are white, like the nostril patch, outer feathers, under-flue and tip of the tail. The female, although bearing the same colours, is far more dull and less conspicuous. The juvenile is very similar to the young red-backed shrike but shows very pale patches on the wing where the white will later appear. The length of an adult is 17 cm (6¾ in).

Habitat
Woodland areas, orchards, olive groves and open ground with scattered trees provide its haunts. It builds a rather clumsy, typical shrike nest in the outer branches of a tree.

Aviary
As for the great grey shrike (page 261), but provide a closed shed-type shelter for winter use, where it can have extra feeding time and a little gentle warmth if necessary. This variety seems to feel the cold a little more than the red-backed shrike.

Food
As for the great grey shrike (page 261).

Breeding
As for the great grey shrike (page 262).

14

Cuckoo, Little Grebe, Dipper, Swallow, Martins, Swift, Moorhen, Lapwing, Nightjar, Partridges, Little Ringed Plover and Sparrows

RATHER THAN HAVE MANY separate groups, I have placed these birds together for convenience, since all are a little out of the ordinary compared with the usual aviary inmate. They have been studied in aviculture and, although the diets are varied, with none apart from the dipper has any difficulty been experienced.

The dippers were hand-reared birds which grew well, and the plumage looked perfect as they reached their self-supporting age; in fact they were as good as any wildling. A little later, however, I despaired of them; whether when at liberty they have a state of untidiness just prior to the moult I do not know but I have never seen one looking quite so like a feather duster as mine did.

Swallows, martins and swifts all prove excellent subjects to hand-rear; so many fall out of nests that I feel sure many others have shared my experiences with these birds, but perhaps not many have wintered them, releasing them as fully adult specimens the following spring.

Plovers and moorhens are interesting to rear and having human beings as foster parents seems to appeal to them far more than to most species. The swallows and martins chattered excitedly in groups upon the arrival of someone with food, but the plovers and moorhens were devoted, almost dog-like, in following me, their little legs working nineteen to the dozen and falling over in the panic to follow and keep up.

Cuckoo (*Cuculus canorus*)[†]

General Description
This completely parasitic bird is a species which, once again, is hardly the pet of an aviculturist. The basic grey-and-white plumage is sufficient to distinguish it from most other species. The underparts are very noticeable, being barred with grey and white, whilst the long graduated tail has white spots. There is also a rarer reddish-brown form, invariably a female, barred with white. The first time I saw this colour form, I mistook it for a mutation, until I was corrected by a real old timer (I wonder if any reader will recall Billy Bell). The feet of the cuckoo are particularly odd: the longest toes, the second and third, face the front, whilst the first and fourth toes face the rear; the cuckoo is the only British bird that I can name in which this occurs. The juvenile specimens can be grey or a reddish-brown with barring all over them. To be truthful, the cuckoo can be a very attractive looking bird, mainly because of its plumage. The length is 33 cm (13 in).

Habitat
Mainly of woodland or spinney, thickets and dense hedgerows, this bird most often deposits its eggs in the nests of dunnocks, reed warblers and meadow pipits, which subsequently become foster parents, although the nests of many other birds are also used.

Aviary
Cuckoos are quite frequently handed in for care; often less-informed people, meaning well, see it as a 'deserted baby bird'. Over the years we must have received nearly 20, some of a very tender age and how these were found baffles me; although many of these same people have offered the opinion 'I think it is a hawk'. Where a young bird has been handed in, it can hardly be thrown out, but be warned, if hand-reared, cuckoos can often take up permanent residence and will still be gaping long after they should have migrated to a far warmer climate. In such cases they must be over-wintered, that bond of friendship must be broken, and the ex-patient must be released when wild insects abound; even then it will be mobbed by most other species. It is far better to be rather brutal, and to feed it only until it is self-supporting and then to release it in July/August of the year in which it was taken, provided, of course, that the plumage is not frayed and it is in good condition; otherwise it will never reach its overseas destination. From my comments the reader will appreciate that the young of these birds become very reluctant to feed themselves if they can get sustenance from a kind-hearted person.

Whilst a novelty for the first few weeks they can become just as much pests to man as they are to their foster parents.

Feeding
A good-quality food base, well mixed with grated carrot and apple, chopped greenfood, hard-boiled egg, minced shrimps and prawns, steamed and minced ox heart and liver and a pinch of Vionate (Squibbs) or similar yeast-based powder, plus mealworms, will ensure that the bird eats all that it really needs. Do not let it deceive you and become a permanent resident.

Breeding
What can I say here, other than to admit I have given it thought in my born-optimistic manner. In saner moments I would be the first to admit that keeping a cuckoo would be quite a challenge but who can say that it cannot be achieved? I would use three large densely planted aviaries with adjoining doors, the outer two housing a pair of dunnocks each and the inner one containing a pair of cuckoos. That would be my first attempt but what timing would be needed?

Little Grebe or Dabchick (*Tachybaptus ruficollis*)[****]

General Description
This bird is a mere 27 cm (10½ in) in length, of dark brownish appearance, with black bill and greenish legs and chestnut hued throat and cheeks. It is a charming and endearing little fellow and the smallest of British grebes, although it seems very seldom to be considered as an aviary inmate.

Habitat
It is quite common throughout most European countries wherever a stretch of water exists; it is found on rivers, streams, ponds etc. but rarely, if ever, is it seen in coastal areas.

Aviary
Our birds originated from some eggs taken from a lad and placed under a broody old game bantam. They came from a nest on a Berkshire stream, which eventually ran into the river Thames at Windsor; in fact we only just prevented the eggs from being blown and added to a collection. It was many years ago, in 1934, when I had these birds. The young were hatched and brooded by the foster parent and she and her

little ones were placed in an empty duck pen; this included a few metres of natural pond, the overflow of a natural spring at the end of what was then our garden. The antics of the bantam caused great amusement to all our neighbours for, when the little grebes went swimming, she used to paddle as far into the water as she dared, clucking and becoming very excited until the ducklings returned to her. She became quite famous to locals, getting only her feet wet, but fretting and clucking and flapping her wings at the pond edge because 'her' young were adventuring so much further than she herself dared. Even though I was only about 11 years old at the time, I recall the enclosure well. It was of a height which allowed my father to stand erect and, since it contained a large old evergreen (probably laurel) and part of the duck pond, which was separated from the rest by wire mesh, it must have been at least 4 m^2 (13 ft^2).

Food
Looking back now after years involved in animal nutrition, I wonder how the young grebes managed to survive – the bantam was given a warm mash morning and night, plus all the leftovers from my father's bird room and aviaries, and our softfood even then included some vegetable matter, grated cheese, maggots and a variety of seeds from canaries and finches – yet live they did, seeming to thrive upon the variety, even eating the remains of all the chickweed.

Breeding
I cannot recall, after so many years, just what happened to the third of these birds, but the remaining two decided to nest, using an old branch laid at water level. They built their nest of chickweed, hay, grass and other growing vegetation, in fact almost everything that we offered. They even used rotting vegetation from beneath the water in their part of the duck pond. Five white eggs were laid but, within a very short spell, they looked anything but white, for they became the same colour as the surrounding nest. To me, at that age, they were the finest little grebes in the world. When we moved to Buckinghamshire, to a house with a far smaller garden, all four of the by-then surviving birds had to be released. My sorrow at having to part with them was great; they were only common little grebes but their confidence in me, and their tameness – they almost ate from my hand – was unforgettable.

Dipper (*Cinclus cinclus*)[†]

General Description
At 18 cm (7 in) long, this bird is easily identifiable when seen in mid-stream on a rock, where it spends much of its time bobbing and curtseying then diving into the fast-running water to appear some distance from where it went under. Shaped rather like a wren, but far larger, it is very dark brown above, with the throat and chest pure white; the belly is rufous, rich in colour but fading to dark brown in the ventral area.

Habitat
It is found near fast-flowing hill and mountain streams where pebbles, never mud, cover the bottom.

Aviary
I have kept a few of these birds, but was never able, at the time, to provide a natural setting of shallow running water over a stone-strewn stream bottom. I would suggest, however, that considerable thought is given to the planning of an aviary, with much research being done on their natural habitat and general needs. A 2– 3 m (6½ – 9¾ ft) stretch of water by 1 or 2 m (3¼ or 6½ ft) wide, re-cycled by a pump, with the introduction of a host of wild water insects, could well make these birds feel more at home. Mine were all hand-reared specimens and took our standard food mixture with the usual additives. After providing them with a daily bath, they would, after a period of months, look a little wet and miserable; they were not unhappy but something was evidently not quite right physically. On the other hand, when offered a bath for short periods, or only twice a week, they looked dry, but still miserable after a few months. Their appetite was good and I varied the food extensively. I believe that the correct aviary conditions could make a vast difference. Give plenty of ornamental rock surroundings. The rear and sides of the aviary, if constructed of brick, could be rock-faced and have nesting sites, holes about 25 cm (10 in) in diameter, built into them; this, covered in ivy, would be ideal and far from unattractive. I would suggest an open-fronted shed-type shelter be attached for their use, when they so wish, and perhaps extra light in winter, but they require no heat.

Food
I gave our standard mixture, with ox heart and liver steamed and minced, fish both raw and steamed and fish roe in its natural state.

Cod-liver oil was added, along with plenty of calcium to replace the crustaceans taken in the wild, plenty of vegetable matter and even increased amounts of seaweed in lieu of algae and also for the extra minerals it provides; many freshwater shrimps were given in shallow dishes but at that time I did not offer small minnow fry. I was not altogether happy with the results. The birds ate well and visitors used to congratulate us upon their appearance but when one has lived for years with birds, eating, sleeping and dreaming them, a niggling feeling that all is not quite as it should be grows until a decision is made to release the subjects and start again another time when one has had more time to study the wild specimen. Perhaps one day another opportunity will arise.

Breeding
The netting of a wide variety of water-life is a simple matter and the dedicated bird keeper will never hesitate if a bird such as the dipper is being kept. The stream, as described previously, must be well stocked and perhaps one or two smaller inlets could be added so that these stillwater sections could encourage more prolific breeding of insects. If the hand-rearing, as I experienced, was so simple, why should adulthood bring problems when normally the reverse occurs; maturity, after rearing on a good basic food, is usually a trouble-free time.

The nesting materials in the wild are grasses, moss and dead leaves; the nest is a large domed structure with the entrance hole pointing to the water just below it. Perhaps I was too self-critical at that time; we sometimes expect too much. Had I kept the birds, instead of releasing them, they may have moulted into perfect adult specimens.

Swallow (*Hirundo rustica*)[†*]

General Description
The long wings and forked trailing tails are probably well known to most people observing wild birds. The swallow has glossy blue-black upper parts and a rich chestnut throat and forehead. The tail, with the long outer streamers, is marked with white and the underparts are white, softly tinted with pink. The juvenile has a noticeably shorter tail fork. The adult length is 19 cm (7½ in).

Habitat
Seldom a city bird, this migrant frequents more open country, where it is to be seen soaring elegantly whilst taking insects on the wing, or

swooping low over water where much of its food is taken from the surface. I had a pair nest over my bed in a Nissen hut when in the army; they swooped in through a broken ventilator to a shelf where I kept a few books and there they nested. Stables, mills, granaries, anywhere with an available entry through an open window or door are the favourite sites; there it will use the upper surface of a beam or rafter, or even a brick ledge.

Aviary

Unless being kept for some sphere of research or special project or being hand-reared if fallen from the nest, any aviary always seems to limit them too greatly. All mine over the years have been hand-reared specimens. Some we have wintered in roomy bird-room flights, liberating them in spring, but, when housed outdoors, the greatest consideration was given to their needs; the maximum area I could allow them was about 9 or 10 m (29½–33 ft) by something like 3 m (9¾ ft) wide. Ours used taut string perches and shelves, and the enclosure had a small pond where they frequently swooped to gather flying mosquitoes. Not once did I see them grounded outdoors although I did see it once indoors when entering the bird-room very early; one was waddling, I can explain it no other way, across the floor.

Food

When these birds are hand-reared, diet is no problem. It is provided by a good basic softfood, a little finely minced dandelion or similar vegetation, steamed and minced ox heart and liver, a few chopped-up mealworms and moth larvae, all well mixed together into a crumbly-moist food. The food receptacle should be placed only just below the perches, not beneath them; it can be some way away, but only a little below their perch level. Their water dish, too, if indoors, should be at least 1 m (3¼ ft) from the floor, and be a large shallow container. They will frequently do a 'belly flop' into this, hardly pausing in their flight but getting sufficiently wet to satisfy their wish for a bath. With regard to these food and water containers, I found that the most convenient were wide and shallow but having a definite upward-pointing rim; mine were the white-enamelled metal-type used for cooking in the oven. Mealworms cut up small, and wood-ant larvae seem favourite items when fed to this bird over a long period. When hand-reared, the swallow will readily adapt to a home-prepared food and I have always endeavoured to keep them on as much softfood as possible, with all insects cut small and mixed well in it, for fear that they might revert quickly if given the opportunity to hawk for livefood. I must admit that this has never happened with those that I have kept; ours invariably took the mixture even after having been flying in the outdoor enclosure and brought in for the winter.

Breeding
The only birds we have wintered were hand-reared specimens: wild adults have always recovered from their accidents prior to the migratory departure time and been released. There have been occasions when young birds handed in to us as casualties have been so badly soiled and ill, through well-meaning people trying to keep them for a short period, that their plumage, even upon becoming self-supporting, was not considered adequately air-worthy for migration. These birds were the ones wintered in the bird room. In early spring they were released into an enclosure which housed waxwings, very docile birds, and all went well until the swallows (there were three on that occasion) began taking nesting materials to a shelf beneath the roofed section. As fast and as hard as they worked, the clumsy waxwings knocked down the nest foundations from the shelf; it had always been one of their favourite roosting or resting sites. The swallows were taking insects on the wing so we liberated them and as it was then only the end of March it is well possible that they bred that year.

House Martin (*Delichon urbica*)[†]

General Description
At one time this bird was extremely common but, nowadays, it is far less so. It cannot be mistaken for a swallow since it lacks the tail streamers; the tail in this instance is a very short fork. The upper parts are a dark blue-black except for the white rump and the underparts are pure white. Juvenile plumage is browner. The adult length is 12.5 cm (5 in). I always feel that swallows and martins have rather pretty faces; the large eyes and dainty bills give them a most attractive look.

Habitat
This migrant will be found near human habitation, using the undereaves of a house to build its nest. In more open terrain it will use outcrops of cliff or a rock face. It is a little more gregarious than the swallow, frequently nesting in large colonies on houses close together. It is most familiar to all as a bird that wheels and dives in its acrobatic flight after insects or when about to enter its nest.

Aviary

Apart from keeping 'fall outs' which have been handed in for safe keeping, and the odd adult or so from accidents, I have never had these birds from choice, for, like the swallow, they seem to suffer from confinement. Once again, when being cared for, they have had as large an enclosure for their needs as could be spared. On one occasion I could only offer an indoor flight of 2 m^2 (6½ ft^2) but even in this very limited area the birds circled and dived, 'belly-flopped' in their water and even landed to bathe daily. The few that I have wintered have been simple to care for since these were hand-reared, but even adults can be persuaded to take a surprising amount of softfood when they realise plenty of insects are mixed in with it.

Breeding

I have had no experience of them breeding other than being aware that two birds had paired and were carrying mud, at which point they were released. Their behaviour at this time was interesting; my whole family, more or less, happened to be in the garden at the time and my wife and daughter each released one or two birds. For what must have been 10 minutes afterwards, the house martins flew low backwards and forwards over our heads in gradually increasing flight distances. They commenced with about 3 or 4 m (9¾ ft or 13 ft) in a to-and-fro movement, gradually increasing this till they were almost out of sight, only to re-appear for a few seconds over our heads before vanishing in the other direction. With a suitable aviary and an acceptable diet, I have always said that any breeding is possible and it may well be so with house martins, but the enclosure would certainly have to be a comfortably spacious one.

Sand Martin (*Riparia riparia*)[†]

General Description

The smallest of this family and probably the one least seen, its tail is only slightly forked and is rounded at the outer limit; the underparts are white, except for a broad brown chest band, and the upper parts are a sooty-brown. The length of this bird is 12.5 cm (5 in).

Habitat

This migrant is found near water in open country, where it excavates a hole in a vertical sand bank or disused gravel pit or quarry.

Aviary
Having hand-reared these birds only when they were brought to me, and housed them in an aviary only for a short period, just long enough to be sure that they were taking wild insects, I am not very familiar with them. They would need a very large enclosure.

Food
For some reason they do not seem to take as much softfood as either the swallow or house martin. They will eat it readily enough when being reared as young but, once fully self-supporting, they tend to scorn the mixture and select from it all the insect life. They would probably eat a large enough amount if it was gradually introduced but I have always liked to see birds eat a softfood with a hearty appetite. Feed as for the swallow (page 270).

Breeding
Breeding this bird would be quite a challenge. If one decided to study these birds closely in confinement, even the presence of a large expanse of water would not encourage all the insect life which they would require; a suitable nesting bank would also be hard to construct and whether, being a colonial nester, more than one pair would be required to 'trigger off' any breeding impulse is difficult to decide.

Swift (*Apus apus*)[†]

General Description
Measuring 16.5 cm (6½ in) in length, this bird has a forked tail but it is far shorter than that of the swallow, with no streamers, and the fork is held very close together. The long wings are scythe-shaped. Predominantly dark brown, the throat shows greenish-white. The aerial performance is so fascinating, as a group wheel and dive with the excited squealing so well known to the villager, that a description seems wholly inadequate to convey its beauty.

Habitat
It is extremely widespread, nesting in cavities of cliffs and buildings in much the same manner as a swallow. A shallow cup is built on some supporting ledge and made from straw, feathers and a wide variety of materials, all collected during the bird's aerial acrobatics and whilst the material itself is air-borne. These materials are stuck together with saliva.

Aviary
On the very few occasions I have hand-reared these birds and kept them until they were completely self-supporting, they were barely in an outdoor aviary for more than 2 or 3 weeks, and I have noticed that they seem to pick up the habit of catching a moving insect even quicker than swallows or martins; the head swivels and, by the time one has pin-pointed what the bird is looking at, it is eaten. They are so quick, even whilst being hand-reared, that the movement of a fly in the vicinity will cause all the heads to swivel, following every movement of the insect. Anyone who has seen this bird will appreciate that a large enclosure would be vital: my indoor flight for a brood of these was only about 2 m (6½ ft) long by 1 m (3¼ ft) wide and they managed to manoeuvre extremely well in this small space but it was obvious that anything small outdoors would be totally inadequate for their needs.

Food
As for the swallow (page 270), but double the mealworm content of the softfood.

Breeding
My knowledge of their breeding habits, limited to observations of wildlings, does not really permit me to advise on the subject. The collection of materials for the swift, the fastest flying of this group of small species, poses immediate difficulties and one would encounter many more. An open-fronted shelter adjoining an exceptionally long outdoor aviary, and a keeper willing and wishing to specialise – a person dedicated to the needs of the birds, with time to spare – may make breeding the swift a possibility. It was not so very many years ago that the successful keeping and breeding of many wild birds would have been doubted and considered well beyond one's wildest dreams.

Moorhen (*Gallinula chloropus*) ****

General Description
This is so common everywhere that I feel sure most people can identify this bird upon sight. The upper parts are almost black but are seen to be a dark brown when handled. The slate-grey underparts have a distinct white line along the flanks, joining the white under-tail coverts. The bill is red at the base and this colour extends to the forehead. The female is more dull in appearance. The length is 33 cm (13 in). It is very territorially inclined as far as other moorhens are concerned.

Habitat
It is found in town or country, wherever there is a stretch of fresh water or swamp, and it feeds upon the grassland adjoining water.

Aviary
Provided that the aviary is at least 4 m (13 ft) long by about 2 m (6½ ft) in both height and width and, most important, a pool of water is supplied, a pair can be kept and studied quite simply. A pair will breed in a small enclosure of this size but, if kept for other than educational purposes where one wishes to watch them at close quarters, they should have a far larger enclosure, consisting of grass and pond only. An open-fronted shed will be used by them during the colder months.

Food
Our young birds were incubator-hatched, then housed under an infrared heater, where they received finely chopped dandelion leaves, comfrey, brassicas, water-cress, chopped mealworms, maggots and crushed turkey starter crumbs. As they grew we gave them the standard softfood mixture, trying always to give a little more vegetation than normal. They grew particularly fond of mealworms and would follow me in dog-fashion wherever I went in the garden. They ran so fast when small that their little legs would trip them up. If I walked at a normal pace the length of the garden path, they would almost all have fallen over somewhere en route, in their efforts to follow as closely as they could.

Breeding
Until I kept these birds I had never realised to the full just how pugnacious they can be towards their own kind with the arrival of spring. Almost as soon as the Christmas season was over so they began in earnest. The two birds which we retained cannot be called a true breeding pair; they used an old mud patch in the garden which we kept topped up more or less daily with the hose. I suppose it would have been in the region of 2 m (6½ ft) in diameter, certainly well under 50 cm (20 in) deep and was situated just to the rear of a lawn behind some bushes; since they were more or less reared by hand, these birds were remarkably tame. Even the specimens we released in a local park had not wanted to stay behind when we walked away. We had selected the two most confiding of them and they roosted that first winter in an old dog-kennel given to me by a neighbour. We kept this beneath the bushes adjoining their 'pond'.

As a child I appreciated their first nesting attempts as it coincided with my birthday and, at that age, conjured up all kinds of thoughts. I was led to expect no successful rearing by my father who said it was quite possible they would not succeed in fertilising any eggs until 2 years old. Seven eggs finally appeared on a typical moorhen nest which had been built on our foundation of old bean sticks, broken, and pushed well into the mud bottom of the pool; we gave them old flag and iris leaves but I think they used as much other plant vegetation as these. Seven young eventually hatched but, within 2 days, three had disappeared without trace; whether one of the crow family (we were living in very rural surroundings at the time) was responsible, I never knew. When we moved from that area to a far more suburban locality we had to release the adult pair and the young; the last I recall seeing of them was the pair sedately walking towards their new pond abode with the four young running behind in much the same way as I remembered the older birds following me the previous year.

Lapwing (*Vanellus vanellus*)

General Description
This bird, with its long curved crest and plaintive call in the fields, should be familiar to all. Seen in their groups, quite large flocks at times, in my own childhood, I cannot imagine them being unknown. At a distance it looks merely black and white but, upon close inspection, it will be seen that the mantle, wings and breast are dark green with a rich metallic purple sheen; the underparts are white, as are the cheeks. During the breeding season they have a black throat which moults to white afterwards. They have a splash of chestnut in the wings and also on the upper and tail coverts. The length is 30 cm (12 in).

Habitat
They frequent fallow and cultivated ground, marshy fields and open moorland and are often seen on mud-flats.

Aviary
A 6 m (19¾ ft) length by 2 m (6½ft), both in height and width, would hold a pair of lapwings. The base should be well-pebbled over two-thirds of the area and a shallow pond and a few sedges and clumps of reed should be provided. To the rear of this, allow grass to grow. At the far rear of the aviary should stand an open-fronted shed and it is here that one provides their food.

Food

To simplify feeding we gave these birds our standard softfood plus all the usual additives. A wide dish held their daily ration and they always finished it apart from a few fine crumbs. When using a basic mixture it is a fairly simple matter to add or exclude certain items as one goes the rounds of the aviaries with the numerous small containers. To feed lapwings on such a rich diet may seem on the lavish side, but they were in excellent condition and bred in the collection to which I gave them.

Breeding

If the aviary is laid out as suggested above, little difficulty should be encountered. The insect content of the food will have to be increased considerably once young are hatched, but the chicks will soon learn that the dish contains what they are looking for and grow quickly, becoming completely independent in about 5 weeks.

Nightjar (*Caprimulgus europaeus*)

General Description

Despite being basically of two colours, grey and brown, this is one of the most beautifully marked species I know; the camouflage provided by the colouring is fantastic and reminiscent of another rarity, the wryneck. The eyes are an outstanding feature and are very large; considering the bird's nocturnal habits, this is only to be expected. It has a fascinating 'churring' call, rather startling at night when heard suddenly from nearby. However, from my own close observations I would say that it is the pattern of the plumage markings which is most striking.

Habitat

It is found in open moorland, commons and scrub areas, where it searches for its food – insects taken on the wing.

Aviary

I have handled only three of these birds in the British Isles, one of which was a corpse, the result of a road fatality.

In 1936 I was handed two nestlings, about 6 days old. A local man, out rabbiting with his ferrets, not too far from Eton College, had found a dead bird, still warm and presumably killed by a ferret. A mere ½ m (1⅝ ft) away lay the two nestlings, on bare ground, with no nest in sight. He knew we looked after all injured birds locally and for that reason brought them to us. Needless to say, the nightjar does not make any nest, but uses only a scrape in the ground, where it lays two creamy white eggs, very delicately marked with dark brown and purple and oddly rounded at both ends; I have seen a number of these on the Continent. My father did comment at the time regarding the wide gapes but could not identify the chicks, not even being sure, from the local's story, of their origin. It was not until the feathers broke through later that the mottled down revealed their true identity. Even in those days, when birds were more plentiful, nightjars never appeared to be by any means common. I do recall my father sounding quite jubilant as he named them, adding that I must release them when they were able to fend for themselves. Presumably the dead bird found so close to them was a parent, although we never saw it. The nestlings would surely have died because, as I learned from watching this species on the European mainland much later, parent nightjars seem always to share the parental duties. During the day one warms the young, while the other does so at night, and so, with a single parent, the young might well have died.

Fortunately the young were hand-reared successfully and came into perfect feather. They ate quite normally of the food provided, frequently during the day, yet at night their movements caused havoc in adjoining aviaries. The enclosure that we gave them must have been in the region of 3 m (9¾ ft) long by 1 m (3¼ ft) wide. They spent most of each day squatting on the large tray which formed their feeding station but, with the coming of dusk, their movements, and perhaps their calls too, alarmed and disturbed many of the other resident birds. As soon as we were fully convinced that there was no loss of body weight, and that they seemed to be taking moths on the wing, they were released within 100 m (330 ft) of the very place where they were found. Perhaps the nocturnal flying and calling would have subsided had we kept them in the aviary and had they become fully adapted to their surroundings. However, I do feel that, unless they form the basis of some very serious form of research involving conservation, or attempts are being made to breed them for re-release into suitable habitats which have become devoid of them, the nightjar is hardly a suitable bird for normal avicultural purposes, despite the unquestionable beauty of its plumage and its fascinating behaviour.

Food
When weaned from the home-made hand-rearing mixture onto a standard insectivore food for adults, the birds were given a greater amount of animal protein, although some fruit and vegetable products were included, minced shrimps and prawns, steamed and minced ox

heart and liver, and steamed fish, plus a number of wax-moths and their larvae and moths taken from the wild by 'tree sugaring' (page 33). This was fed in a very crumbly consistency, because, if allowed to become too damp or sticky, much would be left.

Partridge (*Perdix perdix*)

General Description
It has brown upper parts, with the head, tail and flanks more chestnut-hued; the breast is grey with an upside-down chestnut horseshoe on the lower breast and the legs are blue-grey. It is 30 cm (12 in) in length.

Habitat
It is found on moorland, heaths and arable areas but also in semi-desert.

Aviary
As for the lapwing (page 276).

Food
Partridges eat an enormous amount of vegetable matter, but also take seeds; we offered very little except screenings, but our staple softfood mixture was taken readily at all times, probably because the chicks had been brought up on it. Needless to say we gave extra vegetables, such as brassicas, dandelion, spinach and grated carrot, as well as plenty of animal protein. Maggots seemed always a favourite and, though we rarely gave mealworms, we found that they like young locusts.

Breeding
If the aviary is arranged as for the lapwings (page 276), the tall grass at the rear will serve their purpose for nesting. With the arrival of young increase the livefood supply and give plenty of greenfood.

Red-legged Partridge (*Alectoris rufa*)

General Description
The white throat and black spotting on the upper breast just below the black breast band identify this bird at a glance. This bird is larger than the common variety at 34 cm (13½ in) long. The underparts are chestnut and the bluish-grey flanks show heavier barring in black and chestnut. The bill and legs are red.

Habitat
It has a taste for rather wilder country than the common partridge.

Aviary
As for the lapwing (page 276).

Food
As for the common partridge (page 279).

Breeding
As for the common partridge (page 279).

Little Ringed Plover (*Charadrius dubius*)

General Description
This bird has a narrow white line above the black forehead stripe and a black bill with yellow on the lower mandible. The upper parts are light brown and the underparts are white. When not in breeding condition and showing the black crown and ear coverts, the male is much browner. The sexes are alike but the female's black head markings are browner during the breeding season. The length is 15 cm (6 in).

Habitat
They frequent freshwater shingle slopes and the banks of rivers and gravel pits.

Aviary
As for the lapwing (page 276).

Food
As for the lapwing (page 277).

Breeding
The shingle area of the aviary will be readily accepted once the inmates are sufficiently tame. Any young hatching will need a considerable increase in the livefood, as well as chopped mealworms in the softfood. Offer maggots and very small locusts, with wood-ant larvae and similar food.

House Sparrow (*Passer domesticus*)†*****‡

General Description
When viewed closely, watching a tame specimen, this bird is found to be not unattractively marked. The male has a chestnut mantle and grey crown and rump; the throat is black and the cheeks and underparts are greyish-white. The female lacks the grey crown and black throat but both sexes are darkly streaked on the back in a pleasant pattern. The length is 14.5 cm (5¾ in).

Habitat
Wherever human beings work or dwell so too will this sparrow. Even a cluster of shacks kilometres from anywhere will soon attract this bird.

Aviary
A caught house sparrow is very wild and shy and, in fact, many aviary-bred specimens are just as unruly and nervous. The best method is to hand-rear a nest of them if one wishes to obtain a pair. An aviary for one pair can be the standard minimum-sized enclosure. A wooden nest box, as supplied for the tits, should be placed in position well in advance of the breeding season. Ensure the hole is sufficiently large for entry. The layout of the aviary need be nothing special and can be decorated purely for visual effect; roses are always useful in so far as they attract greenfly, which this bird welcomes when breeding, and hop plants at the rear will also serve this purpose.

Food
I suppose that I have had quite a number of house sparrows over the years. On two occasions I have had quite large collections of the fawn mutation and, at one time, a silver. Our method of feeding was to offer the waste seed from the finch enclosures and, as a basic diet, the same insectivorous mixture that we gave our softbills. If this diet is maintained throughout the year and one commences with hand-reared specimens, the aviary inmates can provide quite a lot of interest and pleasure.

Breeding
When the aviary is laid out as above and nesting materials comprise grasses, moss and feathers, nesting should occur. They will use bits of tissue paper, string and much rubbish when at liberty, but some of these items cause the young to become entangled in the nest.

When young are being fed, they will require plenty of livefood: young locusts, greenfly and moth larvae have been used successfully.

Spanish Sparrow (*Passer hispaniolensis*) ****

General Description
The male of this variety has a crown of chestnut and the cheeks are whiter than those of the British species. The breast is black and the flanks are streaked; the feathers on the back are heavily streaked with black. Both the females and juvenile birds show both the flank streaking and the white cheeks.

Habitat
It is encountered everywhere in Spain, in habitats ranging from wooded areas to cultivated land, especially near rivers and streams.

Aviary
The birds which came into my care, as with numerous others of various species, were the highly valued gifts of a very good friend, another member of ASPEBA, who lived in Spain. Because they were of a nervous disposition and had only arrived from Spain a few weeks previously, no thoughts were given to breeding from these birds. However, they were given an enclosure, about 2.5m (8⅛ ft) long by 1.5 m (4⅞ ft) wide, which contained a rather dense-growing privet hedge; this was well interwoven with hops growing outside but shielding the inmates very well.

Food
We provided both seed and softfood mixtures, plus daily supplies of livefood consisting mainly of mealworms. I knew that a large number of earwigs were present in their enclosure for I had seen them hunting and devouring them; green aphids also abounded since they invariably multiply in vast numbers on the underleaves of the hop. Knowing of their fondness for living close to water and frequenting damp areas, the sparrows had been given an enclosure containing an old earthenware kitchen sink, which held gravel to ensure the water did not become too deep.

Breeding
In 1972 the birds made a very untidy nest which looked remarkably like that of a tree sparrow in every way. It was built high in the privet, about 2 m (6½ ft) from the ground. Since only three young finally emerged and two unhatched eggs were found intact when investigating the nest later, it was presumed that the female laid five eggs only. Those found whole were far too dirty to identify the true colouring; even after

washing they remained very grubby but seemed identical to those of the tree sparrow. The young were reared mainly upon mealworms, apart from the insects that were taken in the enclosure and, for a short period of about 7 days, when we received a supply, some locusts. It was amusing to see the sparrows, in flight, carrying locusts, sometimes almost as long as themselves. Later that year, both the youngsters and parents were passed to an old friend who was also a member of ASPEBA, Herbert Murray of Brentwood, Essex, and I believe that he carried on the strain of these sparrows.

Tree Sparrow (*Passer montanus*)

General Description
Measuring 14 cm (5½ in) in length, it differs from the house sparrow in its chestnut crown, yellowish rump and far smaller throat marking which, in this case, is a mere black bib; there is also a small black cheek patch which is absent in the other bird.

Habitat
Orchards and parkland where scattered trees exist provide favoured haunts. It nests in tree holes, thatched roofs and similar sites.

Aviary
Being a rather shy and retiring bird by nature, it is better to hand-rear a pair than take wildlings. Bearing this in mind one can provide the minimum-sized enclosure so that their behaviour can be observed. Ideally, hops should be grown against a rear wire section.

Food
This is another instance where the waste seed mixture from the finch aviaries can be put to good use. Together with this we have provided a dish of softfood as supplied to the insectivorous species. Screenings when available are eagerly turned over by the tree sparrow, but many insects are taken as well and any chopped mealworms in the softfood are soon removed; locusts at the 24-hour stage seem a favourite; this may well be because they resemble a grasshopper so much.

Breeding
Treat as for the house sparrow (page 281).

15

Corvids, Hawks, Owls and Golden Eagle

MEMBERS OF THE CROW FAMILY make quite intelligent pets; they are well capable of having a sense of humour and can be as mischievous as a kitten. One or two flying loose that have been hand-raised, just so long as the garden is large enough and no neighbours feel belligerently inclined towards them, can give a great deal of pleasure. Never tease them, or allow children to do so, as they are not noted for their patience and, if hurt, can become aggressive in their ways; it is only through such maltreatment that I have ever known one to be spiteful. They can certainly be kept in a large aviary, and be bred from, the diet causing no problems whatsoever.

Hawks are a really specialised subject. I have kept kestrel, sparrow hawk, buzzard and merlin but only the first for any length of time; the others were hand-reared and passed over to other people. I am not in favour of liberating predatory species when hand-reared. It certainly can be done but the training enabling the bird to hunt and live by its catch is a considerably long and involved procedure and requires one to have a close affinity with these birds; too many such birds starve when freed by well-meaning but not fully understanding people. All of these species can be housed in aviaries, long enclosures 6 or 7 m (19¾ or 23 ft) long by 2 m (6½ ft) wide and high, which should have a shelter attached with an entrance hole large enough for the bird to pass through effortlessly when in flight; they will invariably breed in the shelter if a box is secured in position for them.

Owls are birds which are seldom kept just as a pair. Having hand-reared a fair number over the years, I must admit that I am not really very fond of these birds; the only variety which I have kept for any length of time and had young from were scops owls. These were hand-reared like all the others, but when my wife hand-fed eight of them from their fluffy stage when they looked like dandelion heads she fell head over heels in love with them. Whereas hawks require a shed-like shelter with an entrance hole, crows and owls need only an open-fronted type of shelter, which should lead off to a long flight. Apart from the exercise essential to their health, the beauty of the hawks' flight, or of any bird in flight, is one of the great pleasures of keeping and studying them.

Small children ought not to enter an enclosure housing hawks or owls unless they have played an active part in hand-rearing the birds concerned and there has been no break in that contact or relationship. Especially during the breeding season, all these birds can become very territorially minded and some hand-reared specimens will then attack purely from instinct. Others, where a rapport exists, will allow themselves to be lifted from eggs and gently replaced.

Carrion Crow (*Corvus corone corone*)**

General Description
This differs from the rook in having no bare face patch between the eyes and bill. The plumage shows a glossy green sheen, which is not so black as that of the rook, and thigh feathers which fit tightly as compared to those of the rook, which resemble rather baggy plus-fours. This large all-black bird measures about 47 cm (18½ in) but I have seen them larger. Although rather ungainly in movement on the ground, and of heavy flight both taking off and when in the air, it makes up for its clumsiness by becoming very attached to anyone hand-rearing it; the sidling up and nuzzling are all very mercenary of course.

My wife once found a young crow, a 'fall out', whilst out walking with the dogs and the very first thing she could lay her hands on upon returning home, for the poor creature was nearly starving, was some cream crackers; these she quickly dipped in milk and the bird gulped the pieces down as though there was no tomorrow. Every day of that bird's stay with us it had to have some of this food. My wife would place four or five dishes in its cage, one containing softfood and minced ox heart or liver, one holding bread sop, another with mealworms and maggots and so on; the bird would sidle along the dishes, looking in each, and, if there were no cream crackers, it would turn and get a little piece of skin of my wife's hand and twist it and then just stand and look at her, waiting for the hidden dish containing its firm favourite. Fond as I am of this biscuit with cheese, it is not very nutritious for bird feeding and I cannot advocate its use!

Habitat
Farmland, parks, moorland and fields adjoining cultivated land are all frequented and here it finds much of the carrion, vegetable matter, insects and small animals that form its diet. It often takes small game birds on the ground.

Aviary
An aviary of at least 6 m (19¾ ft) in length, with the height and width both being about 2 m (6½ ft) and an open-fronted shed at one end to provide a shelter, will provide all this bird will require. Where the bird roosts or sits outdoors, I would suggest a perch at each end of the enclosure. The immediate underneath should either be cemented to allow daily hosing or have large pebbles to a depth of 30 cm (12 in), which can also be hosed and raked over now and again. The shed also should have a concrete base: if one or two courses of bricks are laid before the shelter is erected on top, the hosing will in no way cause deterioration of the timberwork. The perches should consist of natural branches; try to give them of varying thicknesses, with the bark still on, and renew them once a month. A pond in the aviary will always be much appreciated; this need be just a hollow in the soil with a cement covering, with the water renewed daily.

Food
I have often used our own softfood as a base and added a wide variety of other foods but a more economical manner of feeding is to use turkey crumbs, moistened with milk, into which can then be mixed raw grated root crop – swede, turnip, carrot, parsnip – and greenfood such as grated Brussels sprouts, dandelion, spinach and comfrey. Almost any left-overs from the table can be put through the mincer, and these may include meat and gristle, bacon rind, liver, fish, sausage, any fruit, cheese and rind. Stale milk can always be used on their food, and a little wholemeal bread or cake and broken biscuits given in moderation; the occasional portion of canned dog-food will also be taken. Needless to say maggots and mealworms are accepted at any time. Small fish will be taken from a shallow dish; an angler friend often used to give us a few, up to about 15 cm (6 in), which were soon broken to pieces and eaten.

Breeding
In the shelter at the rear, secure a box about 60 cm^2 (24 in^2) and 15 cm (6 in) deep; cover the base with thin beech or birch twigs and supply many more twigs in the outdoor aviary where the rain can keep them supple. The nesting box should be about 50 cm (20 in) below the shelter

roof, allowing plenty of room for entry when flying straight to the nest from outside. Once the twigs are being carried inside offer plenty of moss, plant stems and dried chickweed, then animal hair, wool and feathers; always cut any wool into small lengths to avoid accidents and entanglement. If a pond has been included in the aviary and the surrounding earth kept damp, they will collect their own mud to assist the construction when binding the twigs and other coarse materials together in the initial stages.

For the successful rearing of young one must increase the animal protein: day-old chicks from the sorting sheds of a hatchery, mice, minced offal, fish and a rabbit cut into small sections can be offered; almost anything is taken at this time. To the basic food add some Abidec (Parke, Davis), calcium lactate, Casilan and Complan (Glaxo).

Jackdaw (*Corvus monedula*)[1*****]

General Description
This is the smallest of our crows, and another bird which makes a wonderful pet. As a child I do not think that I was ever without one or two around the place. I suppose many of them reverted to nature with the arrival of the breeding season but there were always others to take their place. It is black with a grey nape and this silvery-grey contrasts with the glossy-jet black to give the bird a most attractive appearance; the eye too is very intelligent looking. They are well known for the manner in which bright objects catch their eye and the way in which they have often taken and hidden such items. It is about 35 cm (14 in) in length.

Habitat
Towns and villages provide its main haunts and it is often found in the vicinity of very large buildings, such as cathedrals. It nests in tree holes, cavities in rock faces – particularly on the coast, and even buildings where entry may be gained to a well-secluded ledge; it quite often breeds in a colony.

Aviary
As for the crow (page 286).

Food
As for the crow (page 286).

Breeding
As for the crow (page 286) but omit chicks from the diet and substitute mealworms, fully grown locusts and maggots.

Jay (*Garrulus glandarius*)†*****

General Description
Measuring about 34 cm (13½ in) in length, this member of the crow family, although considered a villain, is one of our most colourful species. It is no friend of the game-keeper, or of small birds, whom it robs of eggs and young. The body is rosé-wine-coloured and the black and white feathers on the crown are frequently raised, forming a crest. The wings are bright blue, barred or checked with black, with a white patch between the barring and the wing tip; the tail is black. Although having earned itself a bad name for robbing small birds' nests, the jay also eats a tremendous amount of troublesome insect pests.

Habitat
The jay's favourite haunts are in woodland and forest, but it also frequents orchards and parks. It nests in the fork of a tree very near the bole.

Aviary
As for the crow (page 286).

Food
As for the crow (page 286).

Breeding
As for the crow (page 286).

Magpie (*Pica pica*)†

General Description
This today is one of our worst pests for taking the young of smaller birds. An exercise carried out on three properties, involving over 80 nests, provided us with conclusive evidence of this bird's deeds: not one young bird ever fledged and magpies in large numbers were witnessed time and again over the whole area and were seen actually removing the young in many cases. Measuring 46 cm (18 in) long, the tail being about 25

cm (10 in), it is a striking bird in appearance, with a purple-blue gloss of metallic sheen to the black of the body and greenish gloss to the black graduated tail; the underparts are white and the wing has a large white shoulder patch.

Habitat
A town, village and country bird, it now even enters gardens and one was seen robbing a wren's nest within a few feet of a window in a suburban garden a few doors away from my own home. The numbers being bred, and the amount of food required for the young, gives rise to considerable thought as to how best to control this pest.

Aviary
As for the crow (page 286).

Food
As for the crow (page 286).

Breeding
As for the crow (page 286).

Azure-winged Magpie (*Cyanopica cyanus*)[†]

General Description
The birds which I have kept were all hand-reared specimens, the majority having been reared in twos or threes. However, in one instance, a single bird was reared and became 'imprinted' in no mean fashion, remaining to this day extremely attached to me, but disliking females; on many occasions a mate has been introduced but to no avail. The azure-blue wings and tail, the latter being graduated as in the common magpie, the shining black head and nape, the brownish-grey body and mantle, and the white throat make it most distinctive. The length is 34 cm (13½ in).

Habitat
The open woodlands and olive groves of Spain form its habitat.

Aviary
As for the crow (page 286).

Food
Feed as the crow (page 286), but omit chicks and mice and give more locusts in the fully adult stage, with more beetles, mealworms and maggots.

Breeding
As for the crow (page 286), with food as above.

Nutcracker (*Nucifraga caryocatactus*)

General Description
This large bird, over 28 cm (11 in) in length, is the only one of this size that I know of which is blackish-chocolate-brown yet covered in largish white tear-shaped spots. The white on the tail and under-coverts is very noticeable when in flight and the bill is long and slender. The call is a high-pitched crow-like 'caw' whilst the 'song' is a mere bubbling warble. Resting barely a foot away from them I have never heard more but, if alarmed by sudden movement, it gives a loud rather grating warning cry. If in good plumage, it is an exceedingly attractive bird.

Habitat
It occurs in Scandinavia, the former USSR and the mountainous regions of eastern Europe, in coniferous forests, mainly those with *Arolla* pines. Here it eats pine and spruce seeds, hazel nuts etc., which it tends to store in much the same manner as the jay.

Aviary
Whilst I had these birds, they were housed in an enclosure approximately 5 m (16½ ft) long by 1½ m (4⅞ ft) wide. They were provided with conifer branches at each end of their aviary, with roof cover for 1 m (3¼ ft) at each end. These birds bathed freely in an upturned dustbin lid. No squabbling ever became noticeable, although the male did exhibit the natural dominancy expected at the feeding area.

Food
A mixture of pigeon conditioning seed, to which small pine nuts, sunflower, safflower were added prior to soaking in water for 24 hours, was used. A further dish contained a handful of softfood well mixed with a wide variety of extras, as given to smaller insectivores. Also given was a dry mixture of hazel nuts, acorns, household currants and sultanas, crushed peanuts, and any berries that were available.

Breeding

These birds were added to our collection simply for me to study, since I had watched them in the wild and believed that I could learn much more about them if they were maintained under controlled aviary conditions, when one can almost live with them. Now I am older and wiser, I do wish that I had given them an opportunity to breed. Having seen them breeding naturally and dissected nests, I do know that they use fairly high conifers and construct, with numerous twigs, a thick-based nest, using moss lichen and dry grass to form the cup. Five eggs seemed the normal clutch and these are a pale sea-green, marked quite finely with dull liver-brown. All my birds eventually went to an old friend, who used to contribute much to the organisation of the All-British Bird Show back in the 1950s, but I do not think he ever bred them.

Rook (*Corvus frugilegus*)[†]

General Description
The plumage is all-black with a purplish metallic sheen; the face is grey and devoid of feathers and the thigh feathers are loose. It has a much more stately way of walking than the crow. It is very nearly 50 cm (20 in) in length.

Habitat
Farmlands and fields, cultivated or fallow, where a few tall trees stand will be preferred. It feeds in open country on grain, seeds and insects.

Aviary
As for the crow (page 286).

Food
As for the crow (page 286).

Breeding
As for the crow (page 286).

Golden Eagle (*Aquila chrysaetos*)

General Description
This, as I am sure everyone who has seen it will agree, is a magnificent creature. Blackish-brown when studied closely, in sunlight there is no

mistaking the true golden hues. The head is of a paler shade and when, with the bright sun playing upon it, this great creature turns to look at its keeper, it is easy to realise why poets and writers have often referred to it as regal and king of the skies. The grey tail is barred with brown, the feet are yellow and the legs fully feathered; on occasions when the wings are fully open, they have an awe-inspiring span in excess of 2 m (7 ft).

Habitat
I have watched eagles from a distance on the Continent, yet for all my time spent in Scotland, I have never seen one, other than a captive specimen, within the British Isles. Once so common in England and Wales, but now more or less a vagrant, it still remains a breeding species, but is chiefly found only in the Scottish Highlands and in the Hebrides.

Aviary
Although the eagle is included here, the aviculturist who keeps such birds normally specialises in birds of prey. The bird which I accepted, rather against my will and better judgement, had proven to be rather unsuitable as a service mascot and, in those far-off days, although I felt sure that it might prove unsuitable for me also, I had visions of managing to find a good home for it within a few days. The bird and I developed a rapport and he came to trust me fully and would reciprocate my show of affection by scratching his crown. Eventually a good home was found for him, with the extra bonus of a mate, but he was such an intelligent creature that it was sad parting with him.

Food
Our bird received a sheep's head daily from a local slaughterhouse but, for a change, would get pheasant, partridge, rabbit or, occasionally, day-old chicks, to vary his diet as much as possible. My few friends who enjoyed shooting in those days supplied a very good variety of wild food and, once they had seen and admired him, a number of other people, not known to me, used to call with natural foods of this kind. After parting with this bird, I was informed a few days later, that he was a little overweight but otherwise in truly excellent condition, despite his lack of exercise whilst in my care.

Breeding
I do not feel truly qualified to cover this phase of this bird's life. Visits to zoos and conservation parks would prove a better source of such information. Having watched a number of different eagle species in the

wild, I am aware of nesting requirements up to a point but to write as though fully conversant on this subject would be totally dishonest.

Kestrel (*Falco tinnunculus*)†***

General Description
The male has a chestnut mantle and wings, spotted with black, and a blue-grey head and tail, the latter having a black bar and being tipped with white. The underparts are buffish with black spots. The female lacks the blue-grey of the head and tail and generally tends to be a shade paler. The juveniles resemble the female. The wings of the kestrel are long and pointed; the tail, too, is longish, but the most familiar characteristic is its constant hovering flight while watching the ground below. The male measures about 35 cm (14 in) and the female between 36 and 38 cm (14¼ and 15 in); she is invariably a little larger.

Habitat
It can be found in both wooded and open landscapes.

Aviary
As for the crow (page 286), but the shelter should be of a shed type with open entrance hole.

Food
As for the sparrow hawk (page 294).

Breeding
As for the sparrow hawk (page 294).

Sparrow Hawk (*Accipiter nisus*)†

General Description
In flight this bird is seen to have short rather blunt wing tips. The plumage varies enormously. There are darkish-grey upper parts in the male and russet underparts barred with white. The female has far browner upper parts and is whiter beneath; she also lacks the russet, except for a little on the flanks. Males usually measure in the region of 30–33 cm (12–13 in) long whilst the female is at least 38 cm (15 in) long.

Habitat
Wooded country provides its main haunts.

Aviary
As for the crow (page 286), but with shed-type shelter and large entrance hole.

Food
When keeping hawks, one must eventually contact a hatchery and try to acquire the dead day-old chicks which are always available. If this is impossible, white mice will have to be bred specifically for feeding purposes. When using day-old chicks it is best to remove the yolk sac as the high level of cholesterol may affect the longevity of the subject. As a 'make-shift' meal, one can offer a little very lean minced beef. I tried to introduce fish but my birds would not take it; perhaps others may do so if it is minced with, or inserted into, the carcass of a small animal or bird. Remember that one rat is not equal to its weight in mice; in the first case the bird gets only one set of organs whereas, with the equivalent weight in mice, it gets a number of sets, together with the greater undigested food content.

Breeding
Increase the number of mice and day-old chicks; make an incision into each and put in a drop or two of Abidec (Parke, Davis) with some calcium lactate as well.

Barn Owl (*Tyto alba*)[†]

General Description
The upper parts of this bird are a golden-brown whilst the underparts are white; the back and wings are softly flecked with grey and the face is white and circled by an almost orange-coloured ring. The male will normally measure about 33 cm (13 in) in length and the female is a couple of centimetres longer. Apart from the slightly heavier build of the female, the sexes are alike in appearance.

Habitat
A farm barn, church tower or derelict building can house a pair of these birds. They frequent scattered trees and are ever alert for vermin, which form the bulk of their food.

Aviary
As for the crow (page 286).

Food
Mice and day-old chicks will form the staple diet. Offer also mealworm beetles which have finished egg-laying; mealworms will also be taken and fully grown locusts will be eaten once the taste has been acquired.

Breeding
The nest site should be a half-open-fronted box with wood chippings to a depth of 8 cm (3¼ in). Increase the above foods with the arrival of young.

Eagle Owl (*Bubo bubo*)

General Description
This owl appears blackish-brown from a short distance but browner upon close inspection, tinged with yellow. The underparts are lightish buff, with dark-brown streaking, which is very pronounced on the lower breast. The tail and wings are heavily barred and the yellow eyes are fearless-looking, giving the bird an extremely fierce expression. The length is about 56 cm (22½ in) and the wing span is 84 cm (33 in), less than might be expected. The ear tufts will be raised and lowered for no apparent reason, very often at the sight of food being served but, at other times, the sudden appearance of the keeper.

Habitat
A rarity in the British Isles but widely distributed elsewhere throughout Europe and further afield, it usually haunts forests, and well-forested cliffs and rocky areas.

Aviary
Our pair were a gift, absolutely free of charge as the owner wanted them to go to a good home. Although I enjoyed their company, I felt that I was not doing them justice so gave them to another member of ASPEBA. At that time, he had a huge enclosure for them: the walls, back and two sides were all of rock and of a good height, and there was a pond and running water. It seemed ideal but he failed to breed from them. Having watched their wild flight, I would want to give them a flight at least 6 m (19¾ ft) long by 3 m (9¾ ft) wide and 2.5 m (8⅛ ft) in height, a small pond for bathing, a clear area for feeding and a large natural

hollow bole of a tree, or some other large nesting box with some peat in the base. In the wild they will often just make a scrape with little seclusion, even on an open rock.

Food
Rats, mice, day-old chicks and ducklings or rabbit, plus a generous pinch of Vionate (Squibbs) every other day or so should be given. The vitamin additive can be inserted into a small incision made in one of the dead creatures.

Breeding
A large nest box is often preferable to a hollow log, for one can make a small peep-hole for inspection and so keep an eye on what is happening; this, however, depends to a great extent on the behaviour and nature of the adult pair. The clutch normally consists of two or three white glossy eggs and the incubation period will be just over a month. Once the chicks have hatched, the male will be seen to carry greater quantities of food to the hen, who, in turn, will pass the food onto the chicks. Be prepared for about 10 weeks to elapse before there are any signs of fledging.

Little Owl (*Athene noctua*)[†]

General Description
The upper parts of the little owl are greyish-brown streaked with white and the underparts, although similar, are lighter. It is a rather stout little bird, measuring 20–23 cm (8–9 in) in length. The female is invariably the larger. The large yellow eyes give it a rather fierce expression.

Habitat
The little owl frequents parkland and open country.

Aviary
As for the crow (page 286), but with a shed-type shelter and large entrance hole.

Food
Mice, day-old chicks cut in two, mealworms and beetles will suffice.

Breeding
As for the barn owl (page 295), or place a nest box on the shed floor; it should be a complete box with an entry hole on one side. Food should be as above, but increased.

Scops Owl (*Otus scops*)[1*****]

General Description
Measuring a mere 19 cm (7½ in) long, this is a very attractive little bird with its ear tufts and kitten-like eyes. The beautifully marked and camouflaged plumage makes it almost invisible when standing in front of natural tree bark. There appear to be two phases, brown and grey. It is a far more slender owl than the little owl. The male looks a regular little guardsman in the way he stands so erect and slim. He is more slender than his mate, but one must have both birds for comparison in order to sex them with any degree of certainty. The female has a more rounded look about her and even the head of the male seems smaller, but this is only because of his manner of pulling in all his plumage so tightly. The plumage is patterned very attractively with silver, grey and brown; the wings and tail are barred and the thighs feathered. The call of 'pheu' at fairly short intervals is heard more often as the breeding season approaches.

Habitat
It is found in France, Spain and Italy, in broad-leaved trees, parks, farms, towns and villages.

Aviary
As for the crow (page 286), but with a shed-type shelter and large entrance hole. The nest site will be a box with an 8 cm (3¼ in) hole for entry; the base inside should be covered with moist sawdust or shavings. A natural site of hollow log can be offered but ensure that this is sufficiently large inside for their use, at least 18 cm (7 in) in diameter, and add a layer of sawdust or wood shavings.

Food
Day-old chicks, mice, mealworms and their beetles, and locusts can be given in all sizes.

Breeding
All of our birds have been hand-reared and so were exceedingly tame.

The mating took place at dusk after considerable calling by both sexes. The hen seemed to come off her eggs only with the going down of the sun. In the wild, this owl eats mainly insects; very rarely indeed would it take small mice and never anything as large as a day-old chick.

This is a very different bird to the pygmy owl of Scandinavia and the former USSR, which is very fierce for its size, so much so that one can find the remains of waxwings (*Bombycilla garrulus*) in their nest boxes.

Tawny Owl (*Strix aluco*)****

General Description
About 34 cm (13½ in) long, with a wing span of 56 cm (22½ in) or thereabouts, this bird has both a brown and a grey phase. This colouring is not related to the sex of the bird; the female is slightly larger than the male. The variation in plumage from bird to bird can be quite startling, but generally speaking, they are a rich chestnut-brown with varying shades of brown, fawn, buff and grey. The head seems large yet, when handled, the actual skull is not overly so. The barred wings are beautifully patterned with light and dark brown; a really close look at the patterns is worth the trouble, but take care to avoid the talons. A hand-reared bird is, of course, far tamer: we have had many handed in and many have been content to spend their evenings sitting on the covered arm of a chair, like a cat, having their crown scratched and ruffled. Even so they are not suitable as house pets.

Habitat
This is more of a woodland species. When perched on a protruding branch, standing very close to the bole of the tree, it becomes almost invisible unless silhouetted against the sky. The eerie call, made at night, is not unattractive to those familiar with country noises but often sounds weird to the town-dweller who is just visiting.

Aviary
So many injured birds of this species have been handed in over the years. Most are hacked back to the wild, but sadly there are those, perhaps blind in one eye, or with a badly sagging wing, that must be found good homes. Two such birds, which we kept many years ago, provided us with young every year, which were released, until finally the adult pair went to a private zoo in Surrey. During the time they were with us they were housed in an aviary about 4 m (13 ft) long by 2 m (6½ ft) wide;

this had a small shed attached which the birds used a lot during the day, although they were often to be found sitting in the sun.

Food
Day-old chicks, rats, mice and rabbit cut into pieces of a suitable size were all that these birds ever had. We did add small amounts of Vionate (Squibbs) to their food up to three times a week. Whilst they were rearing young we also added to their food a considerable amount of either grated cuttlefish or calcium lactate which we knew would be passed on to their chicks.

Breeding
No actual nest is made, but a large nest box or deep open-topped box some 34 cm^2 ($13\frac{1}{2} \text{ in}^2$), holding about 6 cm ($2\frac{1}{2}$ in) of peat will often be accepted. Our nesting box was sited in the small shed at eye level. Every one of our hen's clutches numbered four, yet some wild nests that I have seen contained only two eggs. Even larger clutches can be found, but the number of ours never varied. Allow 28–30 days for the incubation period. The female will normally feed the young for about 3 weeks, accepting food from her mate. After this time both parents will be seen to share the rearing. The eggs are white when laid but frequently become very discoloured before the incubation period is completed.

Index of Common Names

Accentor
  Alpine 217–19
  Black-throated 219
  Hedge 220

Bee-eater 252–5
Blackbird 192, 193
Blackcap 138, 139
Bluetail, Red-flanked 209, 210
Bluethroat
  Red-spotted 204, 205
  White-spotted 204, 205
Brambling 67, 77–8
Bullfinch 67, 68, 78–81
  Northern 81–4
Bunting
  Cirl 69–70
  Corn 70–1
  Lapland 73
  Ortolan 73
  Reed 72–3
  Rustic 73
  Snow 73–5
  Yellow 67, 75–6

Canary, Wild *see* Serin
Chaffinch 67, 84–6
  Northern variety 86
  Blue 86
Chat, Black *see* Wheatear, Black
Chiffchaff 139, 140
Crossbill 67
  Common 87–90
  Parrot 91–3
  Scottish 87–90
  Two-barred 93–5
  White-winged *see* Two-barred
Crow, Carrion 285–7

Cuckoo 265, 266

Dabchick *see* Grebe, Little
Dipper 268, 269
Dunnock *see* Accentor, Hedge

Eagle, Golden 291–3

Fieldfare 194
Finch
  Citril 95–7
  Green 67
  Snow 97–9
Firecrest 140
Flycatcher
  Pied 173–6
  Red-breasted 170–2
  Spotted 172, 173
  White-collared 176, 177

Goldcrest 140–2
Goldfinch 67, 99–102
  Northern 103–5
  Siberian *see* Northern
Grebe, Little 266–7
Greenfinch 105–7
Grosbeak
  Pine 108–10
  Scarlet 111–13

Hawfinch 67, 113–16
Hawk, Sparrow 293, 294
Hoopoe 255–7

Jackdaw 287, 288
Jay 288

Kestrel 293

Index of Common Names

Kingfisher 258–61

Lapwing 276, 277
Lark
  Lesser Short-toed 224
  Shore 222, 223
  Short-toed 223, 224
  Sky 225, 226
  Wood 224
Linnet 116–18

Magpie 288, 289
  Azure-winged 289, 290
Martin
  House 271, 272
  Sand 272, 273
Moorhen 274–6

Nightingale 206, 207
Nightjar 277–9
Nutcracker 290, 291
Nuthatch 241–3

Ouzel, Ring 194, 195
Owl
  Barn 294, 295
  Eagle 295, 296
  Little 296, 297
  Scops 297, 298
  Tawny 298, 299

Partridge 279
  Red-legged 279, 280
Pipit
  Meadow 226, 227
  Red-throated 227, 228
  Rock 229
  Tree 230
  Water 229
Plover, Little Ringed 280

Redpoll 67
  Coue's *see* Hoary
  Greenland 121–3
  Hoary 121–3
  Hornemann's 121–3
  Iceland 121–3
  Lesser 118–20

  Mealy 121–3
Redstart 202–4
  Black 200–2
Redwing 195, 196
Reedling, Bearded 186, 187
Robin 211, 212
  Siberian Blue 212, 213
Roller 257, 258
Rook 291
Rosefinch
  Long-tailed 123–5
  Pallas's 125–7
Rubythroat, Siberian 208, 209

Serin 127–9
  Red-fronted 130, 131
Shrike
  Great Grey 261, 262
  Red-backed 262, 263
  Woodchat 263
Siskin 67, 131–4
Skylark 225, 226
Sparrow
  House 281
  Spanish 282, 283
  Tree 283
Sprosser *see* Thrush-Nightingale
Starling 248, 249
  Rose-coloured 249, 250
Stonechat 199, 200
Swallow 269–71
Swift 273, 274

Thrush
  Mistle 196
  Rock 196, 197
  Song 197, 198
Thrush-Nightingale 207, 208
Tit
  Azure 177, 178
  Bearded *see* Reedling, Bearded
  Blue 178, 179
  Coal 179, 180
  Crested 180, 181
  Great 181, 182
  Long-tailed 182, 184
  Marsh 184, 185
  Willow 185

Treecreeper 243, 244
Twite 134, 135
Wagtail
  Grey 231, 232
  Pied 232, 233
  White 233
  Yellow 233, 234
Wallcreeper 244–6
Warbler
  Barred 143–5
  Cetti's 145, 146
  Dartford 146–8
  Fan-tailed 148–50
  Garden 150, 151
  Grasshopper 151–3
  Great Reed 158, 159
  Icterine 153, 154
  Melodious 154, 155
  Orphean 156, 157
  Reed 157, 158
  Sardinian 159, 160
  Sedge 161, 162
  Subalpine 162, 163
  Willow 163, 164
  Wood 164–6
Waxwing 250–2
Wheatear 213–14
  Black 215, 216
  Black-eared 217
  Greenland 214, 215
Whinchat 198, 199
Whitethroat 167
  Lesser 166, 167
Woodpecker
  Great Spotted 236–8
  Lesser Spotted 238, 239
Wren 187–9
Wryneck 239–41

Yellowhammer *see* Bunting, Yellow

Index of Scientific Names

Acanthis
  cannabina 116–18
  flammea
    disruptes 118–20
    flammea 121–3
    islandica 121–3
    rostrata 121–3
  flavirostris 134, 135
  hornemanni
    exilipes 121–3
    hornemanni 121–3
Accipiter nisus 293, 294
Acrocephalus
  arundinaceus 158, 159
  schoenobaenus 161, 162
  scirpaceus 157, 158
Aegithalos caudatus 182–4
Alauda arvensis 225, 226
Alcedo atthis 258–61
Alectoris rufa 279, 280
Anthus
  cervinus 227, 228
  pratensis 226, 227
  spinoletta 229
  trivialis 230
Apus apus 273, 274
Aquila chrysaetos 291–3
Athene noctua 296, 297

Bombycilla garrulus 250–2
Bubo bubo 295, 296

Calcarius lapponicus 73
Calandrella
  cinerea 223, 224
  rufescens 224
Caprimulgus europaeus 277–9

Carduelis
  carduelis 99–102
  major 103–5
  chloris 105–7
  spinus 131–4
Carpodacus
  erythrinus 111–13
  roseus 125–7
Certhia familiaris 243, 244
Cettia cetti 145, 146
Charadrius dubius 280
Cinclus cinclus 268, 269
Cisticola juncidis 148–50
Coccothraustes coccothraustes 113–16
Coracias garrulus 257, 258
Corvus
  corone corone 285–7
  frugilegus 291
  monedula 287, 288
Cuculus canorus 265, 266
Cyanopica cyanus 289, 290

Dendrocopus
  major 236–8
  minor 238, 239
Delichon urbica 271, 272

Emberiza
  calandra 70–1
  cirlus 69–70
  citrinella 75–6
  hortulana 73
  rustica 73
  schoeniclus 72–3
Eremophila alpestris 222, 223
Erithacus rubecula 211, 212

Falco tinnunculus 293
Ficedula
  albicollis 176, 177
  hypoleuca 173–6
  parva 170–2
Fringilla
  coelebs 84–6
  montifringilla 77–8
  teydea 86

Gallinula chloropus 274–6
Garrulus glandarius 288

Hippolais
  icterina 153, 154
  polyglotta 154, 155
Hirundo rustica 269–71

Jynx torquilla 239–41

Lanius
  collurio 262, 263
  excubitor 261, 262
  senator 263
Locustella naevia 151–3
Loxia
  curvirostra 87–90
    scotica 87–90
  leucoptera 93–5
  pytyopsittacus 91–3
Lullula arborea 224, 225
Luscinia
  calliope 208, 209
  cyane 212, 213
  luscinia 207, 208
  megarhynchos 206, 207
  svecica 204, 205
    cyanecula 204, 205

Merops apiaster 252–5
Monticola saxatilis 196, 197
Montifringilla nivalis 97–9
Motacilla
  alba 233
    yarrelli 232, 233
  cinerea 231, 232
  flava 233, 234
Muscicapa striata 172, 173

Nucifraga caryocatactes 290, 291

Oenanthe
  hispanica 217
  oenanthe 213, 214
    leucorrhoa 214, 215
  leucura 215, 216
Otus scops 297, 298

Panurus biarmicus 186, 187
Parus
  ater 179, 180
  caeruleus 178, 179
  cristatus 180, 181
  cyanus 177, 178
  major 181, 182
  montanus 185
  palustris 184, 185
Passer
  domesticus 281
  hispaniolensis 282, 283
  montanus 283
Perdix perdix 279
Phoenicurus
  ochruros 200–2
  phoenicurus 202–4
Phylloscopus
  collybita 139–40
  sibilatrix 164–6
  trochilus 163, 164
Pica pica 288, 289
Pinicola enucleator 108–10
Plectophenax nivalis 73–5
Prunella
  atrogularis 219, 220
  collaris 217–19
  modularis 220
Pyrrhula
  pyrrhula
    pileata 78–81
    pyrrhula 81–4

Regulus
  ignicapillus 140
  regulus 140–2
Riparia riparia 272, 273

Saxicola
  rubetra 198, 199

Saxicola (cont)
   torquata 199, 200
Serinus
   canarius 129
   citrinella 95–7
   pusillus 130, 131
   serinus 127–9
Sitta europea 241–3
Strix aluco 298, 299
Sturnus
   roseus 249, 250
   vulgaris 248, 249
Sylvia
   atricapilla 138, 139
   borin 150, 151
   cantilans 162, 163
   communis 167
   curruca 166, 167
   hortensis 156, 157
   melanocephala 159, 160
   nisoria 143–5
   undata 146–8
Tachybaptus ruficollis 266, 267
Tarsiger cyanurus 209, 210
Tichrodroma murina 244–6
Troglodytes troglodytes 187–9
Turdus
   iliacus 195–6
   merula 192, 193
   philomelos 197, 198
   pilaris 194
   torquatus 194, 195
   viscivorus 196
Tyto alba 294, 295

Upupa epops 255–7
Uragus sibiricus 123–5

Vanellus vanellus 276–7